Isaac Swainson

An Account of Cures by Velnos' Vegetable Syrup,

in disorders, deriving their origin or malignity from scorbutic impurities, or

obstructions in the lymphatic system

Isaac Swainson

An Account of Cures by Velnos' Vegetable Syrup,
in disorders, deriving their origin or malignity from scorbutic impurities, or obstructions in the lymphatic system

ISBN/EAN: 9783337310912

Printed in Europe, USA, Canada, Australia, Japan

Cover: Foto ©berggeist007 / pixelio.de

More available books at **www.hansebooks.com**

A N

ACCOUNT OF CURES

BY

Velnos' Vegetable Syrup,

IN DISORDERS, DERIVING THEIR ORIGIN
OR MALIGNITY FROM

SCORBUTIC IMPURITIES;

OR

OBSTRUCTIONS IN THE LYMPHATIC
SYSTEM.

By ISAAC SWAINSON,

Sole PROPRIETOR of the MEDICINE, and only Suc-
cessor to Mr. De VELNOS;

No. 21, FRITH-STREET, SOHO, LONDON.

Scire poteſtatis herbarum uſumque medendi
Maluit.   VIRG.

LONDON,

PRINTED FOR THE AUTHOR;
AND SOLD BY J. RIDGWAY, No. 1, YORK-STREET,
ST. JAMES's-SQUARE.
MDCCXC.

Preliminary Obſervations.

CANDID and liberal as I have experienced the public, and fortunate as my ſucceſs has been in adminiſtering *the Vegetable Syrup of De Velnos*, there are difficulties in my way, of which I daily feel the inconvenience: and I act againſt prepoſſeſſions, which integrity and merit may not remove. Firſt, The genuine Recipe has ever been a ſecret. Secondly, Its power and efficacy have occaſioned the ſuſpicion of mercury. And, thirdly, It is uſed in ſuch a variety of caſes, as to incur imputations juſtly affixed to univerſal remedies.

I am willing to meet theſe difficulties, or any others, that can be fairly adduced. And if I do not remove them, I muſt ſuffer in the opinion of the intelligent reader; whoſe opinion alone I value.

The reaſon of Mr. De Velnos for preſerving the Recipe a ſecret was—that of private advantage; corroborated by the audacious pretenſions of his perfidious agents; who imitated the preparation, and took out patents for thoſe imitations.

It is known to perſons of the firſt credit and character in this country, that after obtaining full ſatisfaction on the efficacy and excluſive poſſeſſion of the recipe in Frith-ſtreet, I purchaſed it for Four Thouſand Pounds. To ſecure my own property, I was under the diſagreeable neceſſity of expoſing the fraudulent pretenſions of Burrows, Hodſon, Baylis, and Mouldſdale; who vended pernicious

preparations

preparations under fimilar names: and I take every occafion to warn the public, that the genuine Syrup of De
Velnos can be prepared only by me.

It may be lamented, a medicine of fuch reputed efficacy
fhould remain a fecret. And I have received a meffage
by the Solicitor of the Royal College of Phyficians, fig-
nifying their defire, from the general account of its effects,
it might be examined for infertion in the Pharmacopœia.
The College confounded the genuine, with the fpurious
fpecification of Burrows to obtain a patent. "On my
explaining the error to the Solicitor, they did not infert the
fpurious fpecification from the Patent Office; and they
could not apply to me for a recipe which has ever remained
a fecret. The mode to lay it open would be by an application to Parliament, which I have no reafon for making,
befides the public fatisfaction : and Parliamentary finan-
ces are too deeply engaged for political meafures, to allow
any provifions for the public health.

The Vegetable Syrup muft therefore remain under the
difadvantage of fecrecy : balanced by that confidence which
is daily increafing by its effects; and, I hope, not injured
by the pretenfions of its proprietor, to honour and integrity.

But many of the moft powerful medicines, in ordinary
practice, are in effect, *fecrets.* Though the general ingredients of Dr. James's Powder are known, the mode of
preparation is a fecret. The Bark, Caftor Oil, and all the
ftrong and effective chemical medicines, are known to be
fo adulterated and imitated as to be in effect noftrums :
but their general tendency is underftood; and they are
prefcribed.

Indeed, if a phyfician were to prefcribe only thofe medicines whofe ingredients he was acquainted with, he
would nearly annihilate his practice. And yet this is the
pretended objection of fome phyficians to prefcribing
Velnos' Vegetable Syrup. They fay, " We know no-
" thing

" thing of its component principles, and therefore cannot
" answer for its effects!" What do they know of the
component principles of Mercury, Antimony, Opium, and
Hemlock?—Just as much as they do of the Vegetable
Syrup; for chemistry will furnish them with equal information on all these subjects. But physicians never recommend medicines from a knowledge of component principles; they are determined merely by effects. And a physician who would confess himself ignorant of the effects
of Velnos' Vegetable Syrup, would be convicted of a species of ignorance which should disqualify him from practice.
If the effects are salutary, and to be obtained only by ad-
ministering it; of what consequence can it be, either to
the patient or the practitioner, that the ingredients are not
known, or that they are prepared by Mr. Swainson, and
not by unknown laws of nature in gardens or in mines?
This is the reason that physicians and surgeons of the
first character prescribe it; as they do any article in the
Pharmacopœia, of whose effects only they are judges.
Hardly a week has elapsed in this season without my having a patient from Dr. Warren or Mr. John Hunter. Will
it be imagined, that the general nature of the Vegetable
Syrup is not understood by its effects? Or, if its being a
secret medicine necessarily implied dishonour in its preparation, would a man of Dr. Warren's judgment and medical skill, as I am informed he does, give it to his own
children; recommend it, as I know he does, to delicate
female patients of the highest ranks; and consign to me
diseased subjects, where the slightest metallic impositions
would be fatal? I could put similar questions on the credit
of the following physicians, who have sent patients to me:
not in conditions of desperation, or on forlorn hopes;
but with candid opinions of its innocence and efficacy,
and with directions and information which did them honour as practitioners and men.

B 3   Dr.

Dr. Heberden,
Dr. Warren,
Sir William Fordyce,
Dr. Afh,
Dr. Higgins,
Dr. Denman,
Dr. Dale,
Dr. Johnfton,
Dr. Pitcairn,
Dr. Reynolds,
Dr. Sandeman,
Dr. John Grieve,
Dr. Auftin
Dr. M'Donald,
Dr. Carver,
Dr. Black, Edinburgh,
Dr. Ainfley, Kendal,
Dr. Darwin, Derby,
Mr. John Hunter,
Mr. Farquar, Marlborough-ftreet,
Mr. Bromfield, Conduit-ftreet,
Mr. Bromfield, Wardour-ftreet,
Mr. Chandler, Blackfriars-bridge,
Mr. Grindall,
Mr. Hall, Long-acre,
Mr. George Wilfon, Henrietta-ftreet,
Mr. Mortimer, Frith-ftreet,
Mr. Griffiths, Cleveland-row,
Mr. Bryant, Chatham,
Mr. Hawkins, Croydon,
Mr. Dundas, Richmond,
Mr. Partridge, Apothecary, Nottingham.

I have therefore, at this time, to contend only againft
the interefted infinuations of fome low practitioners among
furgeons and apothecaries; who, in knowledge, and the
preparation

preparation of paltry or adulterated drugs, should be ranked with rat-catchers.

It is by the whispers of such reptiles, in credulous and ignorant families, the most improbable of all opinions is in any degree sustained, that the power and efficacy of the medicine are owing to mercury.

It is known at this time, to some thousands in London, that the ingredients of the medicine are very numerous; and from a view of my laboratory, it is seen the process is tedious and laborious. I have made many improvements in that process, from a knowledge of chemistry, which my predecessors did not possess; and the effect is acknowledged in the superior efficacy and elegance of the medicine. To what purpose would be such trouble, if a vehicle only were to be prepared for mercury? And why should mercury, in my hands, cure diseases, which it will not effect in those of others? Dr. Warren and Mr. John Hunter, or any of the physicians I have named, understand the operations of mercury in all its applications. They know, that a random and indiscreet use of it is fatal. And would they consign their most opulent and best patients to that danger, in the use of Velnos' Vegetable Syrup, if they supposed it contained mercury?

But every chemist must, or may know, it does not. Dr. Watson, bishop of Landaff, and Dr. Hinchliffe, bishop of Peterborough, Dr. Ainslie, the late professor of chemistry at Cambridge, Dr. Spencer Madan, prebendary of Peterbo-rough, have recommended the medicine with a warmth which has given offence to interested practitioners; and I might not have taken the liberty of mentioning them, if the pen of satire had not pointed them out in publications of very extensive circulation. It is my duty to attempt doing justice to the motives of their recommendation. A gentleman of high character and learning, in the University of Cambridge, had scorbutic ulcers which, in the com-

mon

mon treatment by mercury, brought on hectic and confumptive fymptoms of the moft fatal kind. The Faculty having taken leave, and the laft ceremonies of religion being in view—the Bifhop of Landaff, wifhing to catch at any hope of faving his friend, did me the honour of writing to me; and, on my opinion, the Syrup was adminiftered; refcued the patient, and reftored him to full health. From that moment, his Lordfhip became the decided and warm friend of the medicine: and I take this public opportunity of bearing my grateful teftimony to the uncommon patience and humanity with which he has received and anfwered the numerous applications and letters fent to him on the fubject of it. I need not inform the reader, that his Lordfhip is a chemift; and that it is not very practicable to pafs a mercurial impofture, for a vegetable preparation, on his numerous and valuable friends.

On his expreffing his furprife at its efficacy, and intimating, that his friends had doubts *which he had not*, I offered to fhew his Lordfhip the recipe: and my confidence in his honour may be judged, when I folemnly declare, I fhould not think myfelf indemnified for publifhing it, by a lefs fum than twenty thoufand pounds. He waved his hand jocofely,—" No, no; I have no doubts on the fub-
" ject; and I will not accept your confidence, left I fhould
" be tempted to quit my bifhoprick, and adminifter the
" medicine."

The folicitude to extend its benefits to the afflicted—by the Bifhop of Peterborough, by Mrs. Hinchliffe, by Dr. Spencer Madan, &c. &c. may furnifh *medical* venom for the fhafts of fatire; but the fuccefs of their humane interpofitions, and the motives impelling them, leave that within the bofom which fatire cannot reach: and I will not difcredit it by my feeble defcription or praife.

Dr. Ainflie not only recommended it to his friends, but in public lectures at Cambridge, beftowed on it the higheft praife as efficacious and vegetable.

These testimonies may be sufficient to remove the effects of interested malignity, on ignorant and credulous minds. But I will add, that one of the first Commoners in Great-Britain, in point of property and talents, I mean William Pulteney, esq; has paid particular attention to the Vegetable Syrup; and, by his desire, Dr. Black of Edinburgh did me the honour of calling on me, and asked several questions, which I *supposed* to be previous to an analization of the medicine; as he took a quantity with him. I also suppose that he was satisfied, as I have since had a civil message from him: and the reader will perceive, by the following cases, Mr. Pulteney omits no opportunity of recommending the medicine.

It is not necessary to inform the learned reader, that since the death of Bergman, Dr. Black stands foremost on the list of philosophic chemists; and that the first discoveries in air were made in Britain by Dr. Black; which have since been expanded with so much eclat by Dr. Priestley, Mr. Cavendish, Monsieur Lavoisire, &c. &c.

But I will relate a presumptive proof, which will carry the force of demonstration.

Some of the cures related in the following pamphlet, were performed on persons whose literary accomplishments, though of the very first order, constitute only their secondary claims to the admiration and attachment of their friends: and that I have been the instrument of saving them, is among the first satisfactions of my life.

Relief in the utmost extremities occasioned by mercury, and administered by a vegetable preparation, directed their attention to the general practice of physick; which they perceived to avoid the use of vegetables.

A society was therefore formed, who proposed, as a general object, to substitute vegetable for metallic remedies, in all the processes where mercury, steel, lead, &c. are used,—and the first trials were made on Velnos' Vegetable Syrup. B 5

The gentlemen who aſſociated were the following :—

Rev. John Calder, D. D. Furnivals'-Inn, now of Croydon.
Thomas Dale, M. D. Union-court, Old Broad-ſtreet.
Rev. David Williams, No. 28, Great Ruſſell-ſtreet, Bloomſbury.
T. W. Whittle, eſq; Sloan-ſtreet, now of Bamff.
John Wildey, eſq; Store-ſtreet, Bedford-ſquare.
James Matthews, eſq; Leadenhall-ſtreet.
John Carr, eſq; Sloane-ſtreet.
Robert Mitchell, eſq; Newman-ſtreet.

Caſes were ſelected of the moſt deſperate kind, in the various afflictions where mercury and the Vegetable Syrup have been long in competition; and the medicine was ordered for the patients by one of the Society.

I ſhould obſerve, that Dr. Dale was invited into the Society, not only to direct the proceſſes, but to aſſiſt in detecting any poſſible deception in the preparation or adminiſtration of the medicine. The Doctor, having ſeen and heard of the effects of the Vegetable Syrup, made no ſcruple in giving his opinion that it contained mercury; but with the liberality of a great and honeſt mind, he now makes no ſcruple to declare he was wholly miſtaken. The effects of their experiments may be ſeen in the caſes, publiſhed on their authority. And they will have the more weight, when it is conſidered, they were undertaken on this condition,—That the Society was at liberty to take any mode of trying the nature, as well as the efficacy, of the Vegetable Syrup; and that if they found proofs from analyſis, reaſons to think, or grounds of ſuſpicion, that mercury or antimony were blended with the Syrup by me, on any occaſions, they would annex theſe proofs, reaſons, or grounds of ſuſpicion to the public caſes.

The

The affiduous and ingenious humanity with which Dr. Dale conducted thefe experiments, greatly engaged the attention of the fociety; and it is to be hoped his found learning, fterling merit, and confcientious practice, will attract from the publick the illuftrious notice they deferve.

To terminate all real grounds of fufpicion on this fubject, I will inftruct any gentleman, or any furgeon or apothecary, acquainted with the elements of chemiftry, in a certain and almoft inftantaneous mode of difcovering the *fmalleft* quantity of metallic preparations intermixed with a vegetable fyrup. I have been obliged to give particular attention to fuch a mode, as the imitations and adulterations of my medicine, by adventurers, agents, &c. are numerous, artful, and perplexing.

The only remaining objection, of any weight, in my knowledge is, that the Vegetable Syrup is recommended for fo many difcafes, as to be fubject to the imputations on *univerfal remedies.* .

The Vegetable Syrup of De Velnos is a remedy only, *where morbid matter has been retained, repelled, or introduced into the lymph.* This effect may take place, by various caufes obftructing perfpiration, by contufions or other injuries; and by the introduction of *virus* of various forts, by means of the abforbent veffels.

When the lymph is rendered impure, the glands are foon affected in the mefentery, in the lungs, in the liver, in the ftomach, in the neck, &c. &c. and diforders of fatal tendency take place, to which phyficians have given various names: but the reader will perceive they are branches from the fame evil root; and that the Vegetable Syrup, while it may cure difeafes fo apparently different, as palfy and confumption; or while the public faith may be ftartled at feeing cafes in dropfy, fcrophula, cancer, rheumatifm, gout, &c. *the medicine is in fact directed only to one object,* that of expelling morbid matter from the lymphatic fyf-

tem;

tem ; or, in language more intelligible, reſtoring the ſalu-
tary purity of all the juices.

I ſhall endeavour to render this fact intelligible to every
attentive reader in the following publication, by tracing the
various diſorders I have occaſion to mention, to a common
origin.

It is by an attention to this circumſtance, that I am
enabled to give advice ; and while common practitioners
are employed on different ſymptoms, I have the general
happineſs of taking away the ſource of various and melancholy evils in the conſtitutions of Engliſhmen.

SCURVY.

Letter from Mr. HEWERDINE to Mr. SWAINSON.

S I R,

GRATITUDE to the means of a cure, of which I had some reasons to despond; and a strong inclination to add my testimony to the high and merited fame of the Vegetable Syrup, induce me to furnish you with the subsequent case.

In the year 1787, I was on the coast of Essex, for the purpose of sea bathing, which in some time produced effects that alarmed me, several eruptions appeared in different parts of my body, which I suspected to be scorbutic; and the suspicion was confirmed by a medical gentleman of eminence; under whose care a cure was attempted by various powerful medicines, among which I had reasons to perceive mercury was not omitted. My eruptions became tumours, and my hopes were yielding to despair, when I applied to you. A course of eight or ten bottles perfectly cured me; and I have not since observed the slightest symptom of scorbutic humour in my system.

I am, Sir,
Your obliged humble servant,
W. HEWERDINE.

Charles-street, Westmin-
ster, Nov. 1789.   CASE

C A S E II.

S C U R V Y.

Mr. FRYER, cabinet-maker and upholsterer, No. 472, Strand, had for six or seven years scorbutic eruptions on the face and ears, which were very troublesome and disagreeable. He consulted gentlemen of the faculty, who proposed and tried several methods of relieving him, and though they disclaimed the use of mercury, administered it secretly, and were detected only by salivation.

When he entered on a course of the Vegetable Syrup, Mr. Swainson apprized him of the difficulty of effecting a cure in the face, by a medicine acting principally by the channels of perspiration.

But Mr. Pryer used the necessary cautions respecting the repelling power of cold; and by taking a considerable quantity of the medicine, was perfectly and permanently cured.

C A S E III.

S C U R V Y.

The Case of the Rev. Mr. GREEN, of Huntingdon.

AMONG the numerous instances of the efficacy of the Vegetable Syrup, the case of Mr. Green, though important, might probably never have been inserted, if Mr. Perry, Surgeon, of Argyle-street, had not observed it;

and

and with that lively zeal which feems to be his difpofition, inferted an account of it in the public papers. It is therefore neceffary for Mr. Swainfon fimply to relate, that when the Rev. Mr. Green applied in Frith-ftreet, he faid he had frequently obferved fcorbutic fymptoms on feveral parts of his body ; that after a fit of the gout, an excoriation of the leg took place; that he had been under the care of Mr. Bromfield, an eminent furgeon, who had treated him with fkill and humanity ; that he had confulted Dr. Warren, who with his ufual judgment thought topical applications alone would not effect a cure; and that he had obtained the full approbation of Dr. Warren and Mr. Bromfield, to enter on a courfe of Velnos' Vegetable Syrup. The effect of the medicine was fo fudden (as is fometimes the cafe) that in fix days the ulcerous appearance of the leg was greatly amended. To be certain that a change fo aftonifhing was effected by the Syrup alone, he omitted it a few days, when the ulcer fhewed evident difpofitions to degenerate into its former condition. On refuming the medicine, found granules of flefh appeared, the wound was healed, and the leg has remained found ever fince.

I had great fatisfaction at the time in refcuing from mifery fo worthy and valuable a man as Mr. Green ; and I acknowledge with thankfulnefs his unwearied affiduity in extending to others the benefit he has received from the medicine.

January, 1790.

LEPROSY.

THE complaints which in this country receive the denomination of leprous, are generally owing to an incompetent perspiration, or to sudden obstructions of it, in the Spring and Autumn. Morbid matter, by these means, is retained in the lymph; conveyed into the blood, and occasions inflammations and obstructions of various appearances.

Inflammations, called scorbutic, may proceed from numerous causes: from blows, contusions, and such external accidents; from the contraction of the smaller vessels, external or internal; from spasm, compression, and obstruction of numerous kinds; from difficulties in the passage of the blood at the extremities of the arteries; from its thickness, viscidity, mixture with heterogeneous or virulent matter; and various other alterations. These alterations generally commence in that liquid called the lymph; which conveys to the blood and habit almost all the good and evil it can receive.

It is by its operation on the lymphatic system, and purifying what are called the juices—that the Vegetable Syrup proves a medicine of such extensive effects.

Under this article, some species of cutaneous eruptions are classed.—And I think it necessary to apprise the attentive reader, that the activity and acrimony of diseased humours are so great, that the most obstinate and dangerous

rous

rous difeafes, both acute and chronical, are terminated by eruptions, and brought back by repelling them. Of this fort are gouts, epilepfies, fpafmodic afthmas, fevers, &c.

In all difeafes of the fkin, even when attended with tumours, I am often teifed with applications for external affiftances, in the ufe of the Vegetable Syrup.

A plain decoction of bran, mixed with new milk, is the beft. Mercurial unguents are the moft dangerous, becaufe the moft powerful repellents. Sulphurous baths may be ufed with the medicine; but a common warm bath, in my experience, has anfwered all their purpofes.

I have feen fatal confequences attend the drying up of pimples and puftules, by ftrong purges; the method of common apothecaries. Lofs of fight has often enfued; and even madnefs in fubjects of great fenfibility. Purges may repel humours from the fkin; but never evacuate them.

I would wifh parents to obferve in time, that maladies of the fkin are often fymptoms of obftructions in the glands beneath it; in the lungs, the vifcera, or the mefentery. The medicine, by its effect on the lymph, refolves thefe obftructions.

I have obferved, in moift and cold fummers, cutaneous diforders have been epidemic. Perfons of dry habits are teifed chiefly by itching eruptions; thofe of fpongy habits, by humid puftules. Some have a malignant kind, with hard tumours under the fkin, and fwellings of the glands; particularly in the arm-pits. Children have deep and obftinate ulcerations; eryfipelas of the head and face; and coughs more or lefs violent, as the eruptions appear or difappear.

I muft charge the reader to remember, that no external remedies be ufed, until the humours are purified by the Syrup. If repelled, the moft violent diforders may take place—as confumptions called galloping; fpafms, vertigoes, epilepfies,

the

the iliac paffion, inflammatory fevers, deliria, palpitations of the heart, afthmas, and dropfies. Thus occafioned, they are extremely difficult of cure; and I believe never cured without the return of the eruption.

C A S E IV.

L E P R O S Y.

Mr. ROBERT HUTTON, at the Penny Poft Office, oppofite Mount-Row, Lambeth-Marfh, was for years af-flicted with a moft violent fcurvy. The eruptions and puftules were confluent, and covered the whole body in fuch a manner, that the diforder was pronounced a leprofy. While an out-patient at St. George's Hofpital, and pro-bably in the ufe of mercurial medicines, his joints, par-ticularly his right ancle, had hard and painful fwellings; his knee was contracted: and a hectic fever, want of ap-petite and reft, almoft brought him to his grave. The firft bottle of Velnos' Vegetable Syrup gave him hope; and his cure was effected in fix weeks.

Witnefses to the cure,

> J. Willis, Mafter of the Thatched Houfe Tavern, St. James's-ftreet.

> R. Sutton, Mafter of the Ladies' Coterie, Albe-marle-ftreet.

C A S E

C A S E   V.

L E P R O S Y.

IN the year 1780, Samuel Pogmore was induced, by the heat of the weather, to go into the water in a state of high perspiration. This occasioned almost immediately a thick eruption over the whole body. A latent disease, of a scorbutic or scrophulous nature, predisposed his constitution to such an eruption, which is a common effect of obstructed perspiration in similar habits. He applied to several medical gentlemen, and took medicines, mercurial or antimonial, without doubt, for they are the nostrums of the faculty. The pimples became tumours, which produced large ulcers in almost every part of his body. The matter discharged was so great in quantity as to stiffen his clothes: it was so fetid and offensive, and wasted his strength so rapidly, that he could derive no consolation but from the near approach of death. In this condition he was recommended to Mr. Swainson, the latter end of the summer 1783, by Mr. Dutton. He is now completely cured, and in perfect health.

SAMUEL POGMORE.

At the Castle and Falcon, Holborn.

Witness,
Benj. Dutton, Bedford-street, Covent-garden.

C A S E

C A S E   VI.

A B S C E S S.

Under the Infpection of the Society defcribed in the Preface, and copied from their Regifter; the Cafes taken down by Dr. Dale.

August 1ft, 1786.

MICHAEL BALLARD, aged twenty-four, by trade a baker, was afflicted two years ago with a difeafe, for which he was under the care of a furgeon. When he thought himfelf well, a hard reddifh tumour, attended with much pain, appeared on the infide of the left thigh, which refifted every application, and at length, after fix months, broke. About twelve months afterwards, another fwelling of the fame kind appeared on the outfide of fame thigh, and another on the outward ancle of the right leg. The two laft tumours became ulcerated in three months. For thefe, a variety of remedies were tried, but in vain ; the patient was in conftant pain ; became very much emaciated and debilitated. He was at laft fo miferable an object, that his life was difpaired of. It was in this ftate he began to take the Syrup, on Friday, July 17, 1786 ; fince that time, he has rapidly mended, and can go any where without crutches, which he could not do for upwards of three months before.

Nov. 17, 1786, Examined by Dr. Dale—almoft well, the ulcers on the left thigh excepted, which continue to difcharge confiderably.

Mr. Swainfon thinks it neceffary, as a warning to thofe who trifle with thefe diforders, to obferve, that after reftoring this patient to good general health, it has not been in his power to cure the abfcefs in his thigh ; and that he fears abfceffes in the deep flefhy parts, if fuffered to form completely, and to run for a confiderable time, are incurable.

CASE

C A S E VII.

L E P R O S Y.

Under the Direction of the same Society, and taken down by Dr. Dale.

August 1, 1786.

SARAH CHESSON, of Princes-ſtreet, Lambeth, aged twenty-two, the wife of a waterman, was afflicted with a leprous eruption at the age of ten, for which ſhe had conſiderable quantities of medicine, which have had the effect of removing it, but it always returned after a few months. She was laſt of all in an hoſpital, from whence ſhe came out apparently well; but in three months it appeared again. The eruption now occupies both arms, and both legs, and ſeveral other parts of her body. Examined, Nov. 7, 1786, almoſt well.

In this Caſe, Dr. Dale deſired the attention of the Society to the facts, that the patient had been cured more than once by the uſe of mercurials, or the common method; and that the diſeaſe returned, often aggravated; as he had found it in his practice. He had no doubt at the examination, Nov. 7, but a few bottles would effect a cure; but he ſuppoſed the diſeaſe would return. The cure was effected by a few bottles, and a twelvemonth after the patient came to Mr. Swainſon with ſpots on her arm, as the Phyſician had foretold; but they were few and mild, and ſhe had never been ſo long free from them before.—They were removed by a few bottles; and a ſmall quantity, Spring and Autumn, will keep the patient clear to the end of life.

CASE

C A S E   VIII.

L E P R O S Y.

Mr. JOHN LANE, master of the Angel-Inn and Livery Stables, Birmingham, was afflicted several years by a most inveterate Leprosy; almost the whole body was covered with pimples and scales: the face, head, arms, and legs, were particularly affected, and the matter or humour it discharged was so thin, hot, and corrosive, as to eat into the flesh with excessive pain. In this deplorable situation, with the usual aggravations of want of appetite and sleep, he was advised to try the Vegetable Syrup of M. de Velnos; and in less than three months he was perfectly cured, to the astonishment of his acquaintance, as well as to his own satisfaction and joy.

Witnesses,

P. F. Bourgeois, Merchant, Birmingham;

Thomas Warren, Printer, ditto.

Mr. Swainson has leave to refer to several persons who have been cured in similar conditions.

THESE diseases generally arise from what is called Scurvy, or impurities in the lymph:—and their solution is by perspiration, by urine, by discharges of blood, and by eruptions on the skin.

These circumstances explain the propriety of seeking relief for them in the Vegetable Syrup; as they are only different denominations of one general disease; to the cure of which the medicine is adapted.

Rheumatism, Gout, Ague, and diseases of that class, are most common after hot summers, when drinking plentifully, even of cool acescent liquors, and exposure to cold night air, impede perspiration, and retain humours rendered acrimonious by heat.

The fits in all these disorders, often go off by plentiful sweats; but they return again; and by a repetition of such operations, the patient is frequently reduced and killed.

The effect of the Vegetable Syrup is to restore the natural state of insensible perspiration; and while it relieves, it refreshes and invigorates.

Children are particularly liable to agues, from crudities in the stomach, deficiency of perspiration, exposing themselves to cold air in the night by throwing off the bed clothes, &c. &c.

Obstructed

Obftrutted perfpiration in Spring and Autumn, firft occafioning rheumatic and gouty inflammations, and the various fevers which go under the denomination of ague, the celebrated Hoffman affirms, (Vol. I. p. 28) " they " give rife to dropfies, cachexies, oedematous tumours, " hectic fever, dry afthma, and jaundice ; in aged per- " fons to comatous and paralytic affections; in the young, " to hypocondriacal diforders; in children, to convul- " fions. In all thofe cafes, upon diffection, the liver, " fpleen, pancreas, and meferaic glands, are found ob- " ftructed or corrupted."

Mr. Swainfon enumerates thefe frightful confequences on the authority of Hoffman; as well to warn his friends to a timely attention, as to obviate the objections of ene- mies to the efficacy of his medicine in numerous diforders. They are all branches of the fame evil; and they are re- moved by the Syrup, merely by clearing the lacteal glands and emunctory veffels, for which it is peculiarly calculated.

In p. 29, Hoffman obferves, " that in the obftructions " above alluded to, recourfe is had, in general practice, to " mercurial and antimonial preparations."

In p. 21, he fays, " what ought to be impreffed on the " confcience of practitioners, and the underftandings of " their patients:—The bark is *fafe* only, when the firft " paffages are fufficiently evacuated; *when the body does* " *not abound with impure juices*; when the vifcera are " unhurt, and there is no danger of an internal inflam- " mation."

CASE

C A S E IX.

R H E U M A T I S M.

MR. THOMAS HOWELL, Lambeth-Marfh, had long a fcorbutic complaint, attended with rheumatic pains, which became at laft a general and confirmed Rheumatifm. The pains in every part of his body were fo excruciating, and he was rendered fo feeble and helplefs, that the affiftance neceffary to turn him in bed, kept him for three months in dreadful agonies. Two gentlemen of the faculty attended him, whom he will have the candour privately to name; but he was fo far from finding relief, that they procured no alleviation of his diforder. He was recommended by Mr. Hutton, at the Penny Poft-Office, to the ufe of Velnos' Vegetable Syrup. The firft bottle brought on a perfpiration, and mitigated the pains; and in lefs than two months he was reftored to perfect health. Mr. Swainfon can refer to a multitude of perfons who have been cured of the Rheumatifm in the fame manner.

Witneffes to the cure:——Jofeph Hopkins, furgeon, No. 85, Compton-ftreet, Soho; Robert Hutton, at the Penny Poft-Office, Lambeth-Marfh.

C A S E X.

G O U T.

To Mr. SWAINSON.

SIR,   Madeira, June 23, 1785.

I cannot deny myfelf the pleafure of informing you, that, until the 17th inftant, I have entirely efcaped the

C   Gout

Gout ever fince Chriftmas, except a few flight fymptoms of it in my right wrift, and two diftinct fits in the laft joint of my left fore finger; each of which lafted but a few days. On the 17th inftant, after a great deal of walking in very bad roads here, I was attacked with a pain in my right ancle, which has confined me five days, and exhaufted itfelf in my right knee, which is now a good deal fwelled, but free from any violent pain. The mildnefs of this fit, and the gentlenefs of the other attack above-mentioned, I afcribe to the ufe of your Syrup, of which I took feven bottles in the Spring. I muft juft add, that until I took your Syrup, I have never miffed a fmart fit of the Gout every Spring fince the year 1768.

I am, Sir,

Your humble fervant,

A. LITTLEJOHN.

C A S E   XI.

RHEUMATIC GOUT AND AGUE.

Mr. THOMAS JOYCE, of Warwick-row, Coventry, in the year 1782, had a violent Rheumatic Gout, and was deprived of the ufe of his limbs, which fome time after he partially recovered, though the Gout was not eradicated. In the beginning of the year 1784, the dif-order put on a complicated appearance, and he was fo far from finding relief from the Materia Medica, that the difeafe acquired greater malignity in proportion to the application of medicines; and in the September of the same

fame year, he was fo reduced as to be incapable of walking without affiftance. When he entered on a courfe
of Velnos' Vegetable Syrup, in September 1784, he had
hard fwellings in various parts of his body, particularly
on the fhin bones; his legs and feet were fwollen; pains
in every part of his body, fpitting of blood, total lofs of
ftrength, depreffion of fpirits, and a voracious appetite,
were alarming fymptoms in his emaciated and declining
ftate. On taking the Syrup of Mr. De Velnos, he
gradually and rapidly got better. In the latter end of
December, he was perfectly reftored to health : and from
being a fkeleton, his perfon affumed a full and lufty
appearance. In February 1785, he took cold, which
terminated in an ague; and in the third fit he took five
fpoonfuls of the Syrup of Mr. de Velnos, which immediately brought on a perfpiration, totally expelled the ague,
which returned no more, and he is now in perfect health.

May, 1785.

C A S E   XII.

RHEUMATISM, GOUT, AND ULCER.

To Mr. SWAINSON.

DEAR SIR,

I was feveral years tormented by a diforder, which the
faculty called fometimes the Rheumatifm, fometimes the
Gout. It appeared to originate in a fcorbutic habit, and
an ulcer formed in my right leg. I was under the care
of feveral phyficians and furgeons; the wound varied in
its appearances, but never thoroughly healed; and I was
fo reduced as to be obliged to walk on crutches. I tried

Margate,

Margate, Bath, and in short every thing the faculty advised. Your Syrup had been frequently recommended, but I could not imagine any single medicine could have removed diforders fo complicated as mine. Defpair, however, obliged me to have recourfe to you; and when you brought a furgeon to examine me, and gave me hopes on his difcouraging report, I attributed it to your humanity, not to your conviction. My fatisfaction and furprife were equal to my defpondence, on finding that by the ufe of the Syrup alone, (for you had ordered all my draughts, falves, and plaifters, to be thrown away) the whole habit of body gradually mended, my gouty and rheumatic vifitants difappeared, and the ulcer foon received a found and radical cure.

I fhould have publifhed my cafe fooner, in juftice to your medicine and your character, as well as humanity to others in fimilar afflictions, if, partly in confequence of having recourfe to you, I had not been embarraffed by a law-fuit with a furgeon who attended me, and who is faid (I hope not with truth) to have been fupported in the action by the Corporation of Surgeons. However that be, I defeated the defign; the caufe was tried before Lord Loughborough on the 30th of June laft; and in the Morning Chronicle of the 5th inftant, you may fee an account of it, as a leffon to the furgeons.

I remain, dear Sir,

Your's fincerely,

JOHN FITZGERALD,
Goldfmith and Jeweller,
No. 23, Lower-Holborn, London.

July 20, 1786.

CASE

CASE XIII.

RHEUMATISM.

At Mr. HALL's, No. 6, Mercer-ftreet.

SIR,   Dec. 18.

I have the pleafure to inform you, that your valuable medicine, the Vegetable Syrup, has made a wonderful alteration in my health, which continues to mend every day. I fhould think myfelf wanting in gratitude, did I not return you my moft unfeigned thanks for your kindnefs, and hope to have it in my power to make you amends for your goodnefs. Even at this inclement feafon, when it could not but have been expected I fhould not have been able to have moved about, I gain ftrength furprifingly, particularly in my feet, which I defpaired of ever recovering again the ufe of. I fhall think myfelf obliged if you will do me the favour to call at my lodgings, No. 6, Mercer-ftreet, when I am certain you will be pleafed to fee every fymptom of returning health. In the mean time, I beg leave to remain,

Sir, your much obliged

Humble fervant,

T. RIDER.

SCROPHULA.

IN another pamphlet, entitled, Hints to Families, &c.
Mr. Swainson has obferved, that his attention has
been principally engaged of late, by the children of perfons
of rank and fortune becoming fcrophulous from injudicious treatment in the meafles and fmall-pox. He has
cafes of this nature, which would form a volume; and
fome of them may be feen in Frith-ftreet. But the im-
putation of Scrophula in a family, being without reafon
deemed injurious, he is much reftrained in his communications on the fubject. His opinion, originally founded
on experience, having been difputed not only in private,
but in reputable publications, he has confulted medical
writers on the fubject, of the firft reputation; and he has
the fanction of their authority.

The number and condition of children, now under Mr.
Swainfon's care, afflicted from thefe caufes with malignant and putrefactive difeafes, flow fevers, foul and fiftu-
lous ulcers, &c. &c. would fhock humanity, and occafion
aftonifhment at the general perfeverance in a pernicious
practice.

The order of veffels compofing the lymphatic fyftem,
is the general agent of good and evil in the human conftitution. Any impediments to the natural functions of
this fyftem prove extremely pernicious, by producing
morbid difeafes of the moft deftructive nature. The impediments are produced by obftruction and abforption;
and their general effects are Scurvy or Scrophula.

Thefe

Thefe difeafes do not feem to be hereditary. Scrophulous children and young people, in my experience, have
been generally delicate, with fine fkins and complexions;
gay, lively, irritable, and with difpofitions to irregularity
and excefs in exercifes and indulgencies. In fuch conftitutions, bad nurfing, the meafles and fmall-pox, fudden
expofures to cold, ftrains, bruifes, obftructions of any natural evacuation, improper diet, inordinate fleep, and
want of exercife, may increafe the tenacity of the lymph,
obftruct the glands, and produce Scrophula.

The firft fymptoms of this difeafe, if approaching the
lungs, are, a hard dry cough, and difficulty of breathing,
on moving bri{kly; if in the fpleen or liver, a fenfe of
pain and uneafinefs in the region; if in the glands of
the mefentery, frequent feats of the complaint, the appetite will vary, the breath will be often offenfive, and the
thirft great; pains will take place in the bowels, the belly
will enlarge, and a fluctuating fever will affect the fkin.
The rickets, white fwellings, tumours on the back of the
head, and under the chin, fwollen lips, eruptions round
the mouth and behind the ears, inflamed eyes and eye-
lafhes, morbid appearances of the fingers and toes, &c.
are indications of Scrophula, which fhould be attended to,
and timely remedies applied. Among thefe remedies,
the Vegetable Syrup of Mr. de Velnos has been lately
drawn into the firft notice, and has been fuccefsful far
beyond any preparations of mercury, antimony, and hemlock, on which the hopes of the common practice are
founded.

In every ftage of this dreadful difeafe, the medicine has
been tried in the laft four years, and under the anxious
infpection of many of the firft and moft refpectable families in the kingdom. Mr. Swainfon is at liberty only to
declare the general refult, refpecting the children of perfons of fafhion. The Syrup has always fucceeded, where

deep

deep abfceffes have not taken place, and the bones have not been highly carious. Cafes of the latter kind have hitherto been only alleviated; the general health of the patients have been reftored; their conftitutions rendered vigorous; and the flighter abfceffes and caries removed; but abfceffes, deep and long formed, and bones highly vitiated, have in fome cafes refifted its force; and Mr. Swainfon is very apprehenfive will continue to refift it.

The following cafes are fair fpecimens of the power of the Medicine in Scrophula, or diforders of a Scrophulous tendency.

C A S E   XIV.

S C R O P H U L A.

To Mr. SWAINSON.

Navigation-Office, Birmingham, May 23, 1786.

SIR,

I fhould be wanting in gratitude, if I were to omit the communication of benefits fimilar to thofe I have derived by your Vegetable Syrup. Having received great benefit from it laft Spring, I recommended it to the parents of a child about twelve years of age, who had, every fpring and fall fince fhe was inoculated for the fmall-pox, been afflicted with fcorbutic or fcrophulous tumours on her face and glands, for which many of the faculty have prefcribed, but to no purpofe. The degree of virulence laft Spring exceeds defcription: whilft in this ftate, fhe began to take your Syrup by two fpoonfuls night and morning; its good effects were foon difcovered, and fhe was perfectly cured,

cured, and reſtored to health by taking two bottles only : and what is more remarkable, not the leaſt ſymptom has this ſeaſon appeared; and ſhe is now a fine, hearty, florid girl. I remain, with gratitude and reſpect,

Your moſt obedient humble ſervant,

JOHN RIDYARD.

─────────────────────

CASE XV.

SCROPHULA,

Under the Direction of the Society abovementioned ; and taken down by Dr. Dale, 14th Auguſt, 1786.

ELIZABETH, daughter of John and Mary Patingell, of Paddington, now at Mrs. Ganer's, No. 9, Mount-ſtreet, about ſeventeen years of age, has been troubled with Scrophulous tumours of the neck and throat five or ſix years; had been two years under the care of Sir William Fordyce, (who with a liberal ſpirit, which does honour to his character, ſtrongly recommended this medi-cine.) About two years ago ſhe was affected with eryſi-pelas; after which, ſeveral of the tumours ſuppurated and broke; ſome of thoſe have been healed up, and others are ſtill in a ſuppurating ſtate. Almoſt all the glands of the neck and throat, which have not yet ſuppurated, are now very much enlarged; are moveable; without pain, and without diſcolouration of the ſkin. Has been taking the Vegetable Syrup about one week; and thinks herſelf better.

Examined, Nov. 1786.—Diſmiſſed cured.

CASE

C A S E   XVI.

S C R O P H U L A,

*Under the Direction of the same Society; and taken down
by Dr. Dale.—28th June, 1786.*

MARY STARLING, daughter of John Starling,
Bricklayer, No. 20, Mary-le-bone-ftreet, Haymarket,
about feven years of age, has been affected with Scrophula
about fix months. She has at prefent a fcrophulous ulcer
on the left cheek; a tumour verging towards fuppuration on the right arm juft below the elbow, with confiderable enlargement, and the arm almoft ufelefs. The left
leg affected in a fimilar manner, juft below the knee.
She has been at St. George's Hofpital, but was refufed to
be admitted, as incurable without the affiftance of fea-
water. Mr. John Hunter, and Mr. Pinkerton, have
ordered fome medicines, but without any relief.
      Examined Nov. 7, 1786, almoft well.
  *The mother has been fince to return thanks to the Society;
the child being cured.*

C A S E   XVII.

Under the fame Direction, &c.

August 29, 1786.

JOHN HOSKINS, aged thirty-eight, was afflicted
about three years fince with an intermitting fever, for
which he was admitted into Haflar Hofpital, being then
in the marines, was near a twelve-month in the hofpital,
from which he was difcharged free from fever, but
troubled

troubled with a very difagreeable fwelling, inflammation, and itching of the left leg. Since that time he has been at various times fo incommoded, as to be unable to perform his bufinefs, (that of a hair-dreffer) for feveral weeks together; the diforder of the leg being accompanied with a confiderable degree of fever. It is now what he calls well, as it is never eafier, but it is ftill inconvenient; feveral puftules being fpread over the calf, and the whole leg conftantly troubled with a heat, and painful itching.

Nov. 7, 1786.———Quite cured.

C A S E   XVIII.

C A R I E S   O F   T H E   B O N E S.

Under the fame Direction, &c.

SAMUEL CHILD, of Hereford, has been afflicted with ulcerations, and a caries of the metacarpal bones of the right hand, for fifteen years. Five pieces of bones have been difcharged through the orifices. On October 26, 1786, when he applied to Mr. Swainfon, his general health was much impaired, and he was very much troubled with a hectic fever; the fores likewife had an offenfive fmell. He had been, previous to this application to Mr. Swainfon, fix months under the care of Mr. Bromfield the Surgeon, who treated him well. He was then perfuaded to apply to Geizler, who took what money he had, and then faid there was no help for him, unlefs he could procure more cafh.

Nov. 7, 1786.—Much better, free from fever; no offenfive fmell from the ulcers.

What

What would have been the event of a full trial of the Syrup in this case, Mr. Swainson will not take upon him to pronounce: as he has yet had no reasons to be sanguine in his hopes, where a Caries of the Bones has been of long continuance.

The patient, while under Mr. Swainson's care and the directions of the Society, shewed his hand to Mr. Pott: who told him his life would be endangered if he did not soon submit to amputation; and the hand was taken off the next day.

The patient waited on the Society, very thankful for the recovery of his health.

C A S E   XIX.

S C R O P H U L A.

Palmal-Buildings, Orchard-street.

ELIZ. SIMMONS, aged eleven years, was admitted two years and a half ago a patient into St. George's Hospital with a scrophulous humour in the left arm, and was discharged in three weeks as incurable, and re-admitted six months after into the same hospital. The surgeons advised amputation, which the mother refusing to agree to, the child was discharged in ten days. In the summer of 1788, she was recommended by Miss Pye, (daughter of the late Rev. Dr. Pye, Conduit-street) to Mr. Swainson, who put her on a course of the Syrup, which produced a considerable discharge from six or seven wounds surrounding the elbow. After a course of three months, the wounds all healed; and she is now perfectly well.

CASE

CASE XX.

SCROPHULA.

FRANCES LOWE, daughter of Mauritius Lowe, History Painter, St. Margaret's Church-yard, Westminster, in the spring of the year 1787, and at the age of five, had a tumour just below the knee, as large as a hen's egg, by which she was confined two months. The tumour was poulticed, and the Vegetable Syrup administered, a considerable discharge of matter ensued; the wound was healed; and the Syrup discontinued, as it appeared too soon. For two new tumours arose, and being brought to suppuration by the medicine and poultices, the wounds were so formidable, that Mr. Cruikshanks, the anatomist, pronounced it a case of extreme danger, and thought nothing but amputation could save the life of the child.

The Vegetable Syrup seemed to make use of the wounds as outlets to all offensive matter in the constitution; for they were not only perfectly healed, but the general health of the child was improved and established. The astonishment of the family at the recovery of the child spread the account of it widely, perhaps with exaggeration. Dr. Dale, to whose ingenuity and humanity no mode of relieving misery is indifferent, expressed a desire to have ocular demonstration of the truth of the case. He examined it with great attention; and with his usual justice and candour, pronounced the cure as perfect and compleat as any he had ever known.

OBSTRUCTION

OBSTRUCTION AND SUPPRESSION

OF

THE MENSES.

———————

THE fuccefs of the Vegetable Syrup in the following cafes, drew enquirers into Frith-ftreet, fo numerous and interefting, that a proper attention to them has often conftituted my whole employment.

The irregularities to which women are fubject, for the greateft part of their lives, from any accident that may obftruct the Menfes—are matters of ferious attention; and the common modes of treating them, are rough, brutal, and dangerous.

It will be feen, though the Vegetable Syrup removes complaints of numerous denominations owing to obftructions of the Menfes, it acts on one general principle,— that of removing impurities in the lymphatic fyftem from the repreffion of excretions.

The effect of fuppreffed Menfes, in my experience, are more numerous and various than I can relate or defcribe. Hyfteric women, and girls on the approach of menftruation, have impediments of fpeech, and fometimes lofe the ufe of it: or if the obftructed matter is tranflated to the head, perturbations of mind, deliria, epilepfy, and madnefs, have been the confequences.

The diforder peculiarly to be attended to by mothers is the *Morbus Virgineus*, or Green Sicknefs; which is an
indifpofition

indifpofition of the whole lymphatic habit, incident to young women from the retention of menftrual matter, or difficulties in its firft appearance. They have appetites for fubftances unfit for food, head-achs, palpitations, and faintings.

In thefe cafes, aftringents of the mildeft kind produce confumption. I never knew a girl efcape death, from the common treatment of apothecaries, by the bark, and what they call ftrengthening medicines.

The Vegetable Syrup gently evacuates the humours, and removes the obftructions to the defired difcharge.

But the moft numerous clafs of patients in this complaint, confifts of women, who are irregular, from the approach of the period when the Menfes difappear; and who have fometimes a deficiency, fometimes an excefs of them.—This period is generally alarming; often fatal. Hyfterics, Convulfions of the Uterus, and menacing fymptoms in the head, are its general attendants. It is when fpots and ulcers take place on the fuppreffion, or on the firft efforts of the Menfes to appear, that patients commonly apply to me. And the Vegetable Syrup has not yet failed, in gently inducing their appearance in young people; rendering difcharges regular; and guarding againft the fatal confequences of their difappearance.

This will not appear improbable to the reader, if he recollects, that the retention of menftrual matter in the lymph or fyftem of juices, has precifely the effects produced by any other morbid fubftance called fcorbutic, fcrophulous, miliary, or variolous. It occafions fever, feizes the glands, and produces internal wafte, hectic, and confumption; it flies to the head, occafions epilepfy, apoplexy, or palfy; or it breaks out in tumours, ulcers, and abfceffes.

The Vegetable Syrup meets it in the lymphatic fyftem, as it does any other virus or morbid matter; carries it off

at

at the evacuating veffels; and the diforders difappear, of which it is the origin.

As the treatment in all cafes of this nature is peculiar, I fhall fave trouble to the afflicted and myfelf, by mentioning the peculiarities.

When any difeafe is determined to be a fymptom of obftructed Menfes; or when menftruation is known to be obftructed—the Vegetable Syrup is to be given in dofes carefully regulated by the printed directions: the furface of the body to be kept moderately warm, even to the extremities of the fingers; and the feet to be bathed once a day in warm water—until menftruation takes place; then the medicine and bathing to be difcontinued, but the body ftill kept warm. When the difcharge is over, the medicine is to be refumed, and continued, liable to the interruption of difcharges, until the health of the patient is fully reftored.

C A S E   XXI.

SUPPRESSION OF THE MENSES.

MARY LEWELYN, late a fervant to Mrs. Farren, Monmouth-ftreet, Bath, took a violent cold in December 1784, which occafioned a total fuppreffion of the Menfes, attended by oedematous fwellings of the legs and feet. Indurations appeared in various parts of the body, refembling boils covered with leprous fcales, which fell off, and were fuccceded by others. Her appetite and ftrength declined, and her condition became extremely wretched. She had been attended by a phyfician, and by two eminent furgeons at Bath; and fhe had tried the waters with no beneficial effect. In this miferable ftate fhe was

recommended

recommended to the Infirmary at Briftol, as to the laft afylum of defpairing wretchednefs, when fhe was fortunately advifed to try the Vegetable Syrup of Mr. De Velnos. In fix days it brought on the periodical difcharge, and in three weeks perfectly completed a cure.

Farther information of this cafe may be had of Mr. Pine, printer, at Briftol.

C A S E   XXII.

SUPPRESSION OF THE MENSES.

To Mr. SWAINSON.

Sir,

I owe my life to your humanity, and the virtues of the Syrup you prepare; and I fubmit my cafe to public notice, in hopes that thofe in fimilar circumftances may feek your affiftance.

On the 17th of May, 1785, I was taken ill, in confequence of a cold. I had pains in my ftomach, giddinefs in my head, and fhivering fits. In about a week my legs and thighs began to fwell; and I gradually grew fo big, that I could not ftoop or walk. I fent for an apothecary, who bled me, gave me draughts, and fome dietdrink; but I did not get better under his care; and I had recourfe to Dr. Meyerbach, who told me I fhould foon find relief, and gave me powders and various other drugs; but I grew worfe; the fwelling increafed; my legs burft, and great quantities of a thin watery fubftance ran from them. At this time the menfes were quite ftopped; a warm and bitter water kept conftantly running from my mouth; fo that I could not lie down without

out danger of fuffocation ; the fwelling increafed, particularly about my loins, abdomen, and ftomach ; a general
forenefs, like an inflammation, prevailed over my body,
attended with excruciating pains ; and I was blind for
more than a week. I then fent to Sir John Eliot, and
was under his care a great while, but without hope either
of a cure, or of life. When he flackened his vifits, as if
to avoid the importunities of defpair, and all my friends
expected my death, I was fortunately recommended to
you, Sir, and on the 17th of September, you had the
goodnefs to put me on a courfe of Velnos' Vegetable
Syrup. In three weeks I was fo much relieved by the
prodigious difcharge it occafioned from the legs, that I
could fit and lie down without danger. Sir John Elliot
called as he paffed, expreffed his aftonifhment at my being
alive and better ; and attributing it to his laft prefcription, which I had long difcontinued, he began to exult,
and faid I was now out of danger, and fhould foon do
well. I told him what he faw was the effect of Velnos'
Vegetable Syrup. Aye, faid he ; looking as if I had hit
him in the face ; and fnatching up his hat, he hurried
with precipitation out of the houfe.—I perfevered in the
ufe of the Syrup, gradually getting relief ; but the obftruction of the Menfes did not give way till the month
of January, 1786. From that time I got better with
great rapidity ; and I now, thank GOD, enjoy my ufual
health.

I am, Sir, with great gratitude,

Your obliged and humble fervant,

MARY ASHLEY,

June 14, 1786.   Church-lane, Chelfea.

Witneffes,

Witneffes,

Mr. Maton, Queen Elm, Chelfea.

Mr. Montellier, ditto.

Mr. Thomas Afhley, Church-lane, Chelfea.

Mr. Wakefield, Crofs-lane, Long-Acre.

Mr. Bedford, Coach-maker, Long-Acre.

Mr. Deyken, Long-Acre.

Mr. Tucker, ditto.

Mr. Haynes, New Tothill-ftreet, Weftminfter.

Mr. Jefhies, Wild-paffage, Drury-lane.

DROPSY.

D R O P S Y.

ALL inflammations, if they continue long, are followed by suppuration, and the production of morbid matter; if they take place in the mesentery, they occasion strong obstructions, and produce a Dropsy.

The Vegetable Syrup, by evacuating the morbid matter and removing the obstructions—does not multiply its pretensions.

The reader will recollect, that acrimonious humours will have various external effects, according to the internal parts on which they affix : and that the Vegetable Syrup has but one property,—that of carrying out of the constitution the impurities of the internal juices.

C A S E XXIII.

D R O P S Y.

To Mr. SWAINSON.

Lichfield-street, St. Martin's-Lane,
SIR,   *Dec.* 6, 1785.

HARDLY any circumstance in my life has surprised me so much as the recovery of Sion Girney. When I was ordered by the Insurance Office to rebuild your Laboratory, and sent him to assist, I was actuated more by my own feelings for his situation, than any opinion he could render much service ; and when I withdrew him, it was with a view to afford him a little assistance, in alleviating

viating the misery in which I suppofed he muft foon have gone out of the world.

He was fo fwollen, that his figure and countenance hardly bore a human appearance; and his thighs and legs were covered with fcorbutic fcales. I directed him to apply to Mr. Pinkftan, in St. Alban-ftreet; and he had the affiftance of Mr. French, apothecary, in Coventry-ftreet. The gentlemen, I have no doubt, treated him with fkill and humanity. He was ordered iffues in both legs, and other proceedings were had; but the diforder, inftead of being checked, daily gained on his conftitution, when he fortunately became the fubject of converfation between you and me.

I am not fond of appearing publickly as a writer, even of a fhort letter; but I thought the generofity and compaffion with which you gratuitoufly undertook the cure of a poor man in fuch circumftances, almoft as extraordinary as the effect of your wonderful medicine; and I think it my duty to bear my teftimony to both.

On taking the firft bottle, the watery humour was forced down in fuch a quantity, as daily to fill his fhoes. As the fwelling gave way, the leprous appearances of the extremities were gradually difpofed to heal; and in lefs than a month, to the afto..ifhment of all who had feen him, the dropfy, the fcorbutic fcales, and even the iffues on his legs, all difappeared.

> I am, with great regard, both for your
> humanity and fkill,
> Your moft obedient humble fervant,
> ALEX. CAMPBELL.

Mr Swainfon has feveral other cafes of a fimilar nature, under his care; but they are not ready for publication; and be makes it an inviolable rule, never to infert a circumftance or a hint, which is not ftrictly true, and fully attefted.

SMALL.

SMALL-POX.

THE reader will obferve, that I do not wifh to involve myfelf in the bufinefs of Inoculation, or the treatment of the Small-Pox. I have already more fubjects of attention than are confident with any hopes of quiet and leifure.

My *Hints to Families,* on the prefent management of this difeafe, has occafioned a general alarm ; and to fave the trouble of numerous applications at my houfe, I will introduce here a few explanatory obfervations and cafes.

When the matter of the Small-Pox has taken effect— the objects of every reafonable practitioner are, to correct the acrimony of the humours; to promote eruption and fuppuration, by gentle diaphoretics, if nature be languid ; to refift the tendency to putrefaction in the decline of the difeafe; and to carry off the remaining impurities. It would drive an apothecary to diftraction, whofe hopes of fubfiftence are on the number, and counteraction of his intricate preparations, if I were to affure him, on the general principle already ftated, that the Vegetable Syrup would anfwer all thefe purpofes. But I want not to enter into competition with apothecaries on this ground— " Live, and let live," is a moral maxim. But I cannot behold with indifference, the daily victims brought to me, of the cool repelling practice now in fathion in the Small-Pox.

The

The skin, at the time of the disorder, is exquisitely sensible to the slightest impressions of damp or cold; and the consequences of repulsion, while it preserves the skin, are ·foul fistulous ulcers, extremely difficult to cure; weaknesses or contractions of the hands and feet; tumours near the joints, which degenerate into malignant and wasting fistulas: various disorders of the glands, which terminate in hectic consumption or dropsy.

On the most minute examination of these cases, I have found that the patients have been exposed, after the infection, in cold windy weather; and the evacuation of the morbid matter checked.

I need not observe, that a similar treatment in the Measles will have similar effects. I have had great numbers of children, whose spots have been suffered to disappear from slight cold; whose coughs have indicated putrescent vomicæ on the lungs, and who have had dropsical symptoms from obstructions of the meseraic glands.

I shall only observe, that in the Small-Pox and Measles, when the Vegetable Syrup has been administered, it has assisted difficult eruptions; and no person taking it, has had the usual symptoms, in any great degree, of frightful dreams, oppressions of the breast, pain at the throat, constipation of the belly, or epileptic fits. It corrected the acrimonious matter; and promoted its expulsion, in a manner as favourable to the skin as its exposure to cold, without any of the inconveniences or dangers.

CASE

C A S E   XXIV.

CONSEQUENCES OF THE SMALL-POX.

To Mr. S W A I N S O N.

Sir,

GRATITUDE and juſtice demand that I ſhould ſend you the following account of the extraordinary effect of Velnos' Vegetable Syrup on a child of mine, and eſpecially as it will be the means of adding one more teſtimony to the virtues of that excellent medicine, and may induce others to try it in eaſes where it has not yet been uſed; and I heartily wiſh, for the good of mankind, that it may become an univerſal panacea.

On the 16th of May laſt, my ſon, three years old, had an eruption upon him, but did not appear to be otherwiſe diſordered. As he had been inoculated at ten months old for the Small-Pox, I took this to be either the chicken or ſwine pox, and ſent him out in the air; which being very cold for the ſeaſon, ſtruck the eruption in, as I imagined; for when he came home, the puſtules appeared to be ſunk, and the child was ſoon after very ill with a high fever. He continued in this way for ſeveral days, and was attended by a phyſician and an apothecary, both eminent in their profeſſion; but notwithſtanding all that they could do, by bleeding, bliſters, and other outward and inward applications, he continued to grow worſe; and at laſt had every ſymptom of approaching death; and was given up by the phyſician, and every body elſe who ſaw him.

In this dilemma, anxious for the life of my only boy, I was conſidering if there were any other means to be tried in order to reſtore him, when fortunately I recollected Velnos' Vegetable Syrup, and propoſed to Mrs. Humphrey

to

to give it the child; but she was fearful of trying a medicine, the effect of which, in such a disorder, according
to the cases published, had not been experienced: but I
observed, that if its salutary effects were so very extraordinary in other disorders arising from foulness of blood,
and consequently the regular functions of nature being
thereby obstructed, why might it not do good in this case,
which answered such description? We consulted some
friends, and it was at last agreed, as I had the pleasure of
being known to you, that I should ask your opinion
on the subject: I accordingly waited on you, when you
were so kind as to inform me, that you had never
known the Syrup to be tried in such a case, but assured
me, that if I was disposed to give it the child, if it did
him no good, it would do him no harm.

On this I determined to try it, and accordingly took
home with me a bottle of the Syrup, and agreeable to
your directions, gave the child a table-spoonful thereof,
which was to be repeated every eight hours. The first
dose was given at ten at night, on Friday the 27th of
May; about four the next morning we perceived the
child's fever to abate, and a gentle perspiration came on,
Happy to find such a pleasing change, continued to supply
him regularly with the Syrup, at first as above directed,
and after at longer intervals, as you advised; the child
gradually mended, and before the bottle was quite out,
perfectly recovered, and now enjoys a good state of health.

I am, respectfully, Sir,

Your obliged, and very humble servant,

GEORGE HUMPHREY,

Printseller, and Dealer in Natural Curiosities,
Surry-side, Black-Friar's Bridge.

July 9, 1785.

D   P. S.

P. S.   I have enclosed my child's case, drawn up more at large, which I beg may be laid up in your repofitory for fuch things, as a memorial of this extraordinary cure, and for the information of fuch as may wifh to know further particulars thereof.

CASE XXV.

CONSEQUENCES OF THE SMALL-POX.

To Mr. SWAINSON.

SIR,

MISS C———, of E———— in Devonfhire, having a fcrophulous fwelling of the glands, was to go through I do not know what courfe of antimonial medicines of Meyerfbach's; I heard of it in time, and recommending your Vegetable Syrup as more innocent, her mother refolved fhe fhould rather go through this than the other courfe. She has taken the medicine; the glandular fwelling has wholly difappeared; fhe is perfectly cured; eats with an excellent appetite; and feems, as it were, to enjoy perfect health for the firft time in her life, having from her infancy always been of a very delicate, fickly habit.

Mr. and Mrs. Thompfon, No. 135, oppofite Doyley's Warehoufe, in the Strand, have feen the cure, and will at any time atteft it.

R. B. RASPE.

London, July 5, 1787.

CASE.

CASE XXVI.

CONSEQUENCES OF THE SMALL-POX.

To Mr. SWAINSON.

Lime-Wharf, Birmingham, June 4, 1787.

SIR,

A Child of mine, who is now about thirteen years of age, after an inoculation for the Small-Pox, had, particularly every spring and fall, violent eruptions in her face, and very much inflamed eyes, with white specks on them; in the spring of 1785, she was attacked with a degree of virulence not easily described, which could net be removed by any medical assistance to be procured; when a friend who had received great benefit in a similar case, by taking your Vegetable Syrup, called upon me, and happening to see her, recommended it. I accordingly applied to Messrs. Pearson and Rollason, and can with gratitude and pleasure say, that by taking two bottles, which she did by doses of two spoonfuls each, she is radically cured; nor has a pimple since appeared.

You should in gratitude, and for the good of others, have heard from me before this time, but was afraid of a relapse the last spring and autumn, and this spring; but no such has, or is likely to appear. She has ever since been remarkably healthy.

I am, Sir,

Your grateful and obliged servant,

THOMAS ELWELL.

Witnesses to the cure,
   Thomas Danks,
   W. Felkin.

CON.

THE two following cafes in Confumption are felected out of hundreds in the author's poffeffion; as they are completely fatisfactory, and are related by men of high eminence, both in abilities and character.

Captain Dorfet, the author of *The Principles of Defen-five War*, and of many admired political works, I know only by the intercourfe occafioned in adminiftering the medicine, and by the public opinion, which is extremely favourable ! Of the Rev. David Williams, I need not fay to any who have flightly attended to him, that his integrity is as uncomplying as his abilities are fplendid; thofe very few only who have his intimacy, know the value of his friendfhip, or the virtues of his heart.—The reader will judge, that I have reafon to be fatisfied for having reftored fuch men, from conditions of mifery and hopelefs defpondence, to a very defirable ftate of health.

It fhould be obferved, that they were not only prefcribed for, but vifited in a friendly manner, by the firft medical practitioners of the time, of whom they fpeak with refpect, though their fkill to them was ineffectual.

The laft of thefe cafes has been completed above three years; and from the zealous and effectual recommendations of the patients, I have had more confumptive than any other cafes under my confideration.

It was the opinion of Dr. Cheyne, that Confumptions always proceeded from Scurvy; and the practice, fortune,

and

and fame of the late Sir Richard Jebb, whose utmost skill was employed on one of the patients abovementioned, were founded on that opinion.

I have confulted, with anxious attention, the beft writers on the fubjeƈt; and I find them unanimous in deriving Confumptions from fchirrous, ulcer, or abfcefs in the lungs; thefe, owing to fcurvy, or to mercury, depraving the nutritive juices. Schirrous tubercles on the lungs are fometimes thrown out by coughing.

In children, who conftitute the moft numerous claffes of my patients, an atrophy, or an emaciated ftate, wearing the appearance of Confumption, generally arife from a fcorbutic, or fchirrous induration of the meferaic glands. In thefe cafes, the belly is inflated, and all the funƈtions of the child depraved. I can refer to children, with the countenances of milk-maids, who were brought to me with flaccid and emaciated limbs; the abdomen diftended; the breathing fhort and difficult; the body weak and weary; the bowels fometimes loofe, and fometimes bound; the appetite irregular; cold food coveted; the temples collapfed; the whole face ghaftly; the fhoulders prominent like wings; the food paffed undigefted; pains at the region of the navel; the urine fometimes thick, fometimes high-coloured; and at night, a heat and thirft almoft intolerable.

Learning that, on diffeƈtion, the meferaic glands have been obftruƈted, where the rickets have proved fatal, I have applied the Medicine with fuccefs, where the feveral parts of the children's bodies have been difproportioned.

When Dropfy, Confumption, Afthma, Heƈtic, or Diarrhœa, fupervene, in thefe cafes—they are generally fatal.

Difeafes of a fimilar nature in youth, and particularly in young women on the point of menftruation, are generally owing to going out too foon after the Small-Pox or Meafles, to the repulfion of cutaneous eruptions, or the

D 3   retenfion

retention of menftrual matter, which the conftitution has not fufficient force to difcharge. A fpecies of Afthma, called tightnefs on the breaft, attended with a teifing and dangerous cough; and what is called the hooping-cough, are owing fometimes to an ill conftitution of the air; but generally to fudden obftructions of perfpiration, to the improper treatment of crifes in the Small-Pox, and other morbid diforders, or to the repulfion of eryfipelatous and other eruptions.

Cutaneous efflorefcences, and eruptions to which young people with confumptive tendencies are very liable, running of the ears, and inflammations of the eyes, arife fometimes from a voracity incident to fuch cafes; but generally from an indifpofition of the humours. It is feen, in the following cafe of Mr. Williams, that he was two years in hourly apprehenfion of blindnefs. The medicine acts, in all thefe fymptoms, on one uniform principle. It corrects the acrid ferum; procures a derivation from the difeafed parts to other emunctories; and ftrengthens the relaxed glands.—Setons, iffues, blifters, are troublefome, precarious, and fometimes dangerous.

It will be perceived, by the following cafes, that epilepfy may attend the fatal periods of Confumption. It generally arifes from irritations of the ftomach, inteftines, and nerves; from the remains of fmall-pox, petecheal fevers, healing up of old ulcers, the repreffion of eruptions, &c. for the difeafe is often carried off by eruption, the appearance of the menfes, fmall-pox, meafles, &c.

Chronical epilepfies are generally adjudged to be occafioned by acrid fcorbutic lymph; and they are commonly treated with blifters and iffues.

The Afthma, and palpitation of the heart—fo diftreffing to patients of this kind, I muft mention—as by taking out the roots, all the evil branches have difappeared in the ufe of the Vegetable Syrup.

If

If the palpitation of the heart has arifen from polypous concretions completely formed, the medicine has not cured; tho' it has relieved, by refolving the juices, and rendering the excretions free. If from menftrual or hæmorrhoidal obftructions, or the repulfion of cutaneous eruptions, it has relieved, with the affiftance of warm baths for the feet.

Afthma is nearly related to Confumptions: and in all the cafes I have known, it has been owing to the fuppreffion of acrid ferum at fome emunctories; producing a tenfion of the membranes, which inveft the cells of the lungs, of the arms, fhoulders, back; and at length occafioning a palfy of the parts. On the moft attentive examination of the cafes that have occurred in Frith-ftreet, all afthmatic complaints have arifen from a fubtle, acrimonious, cauftic matter; either external, as metallic vapour; or internal, as fuppreffed fcorbutic perfpiration, or fome virus or morbid matter retained and locked up in the habit by repelling circumftances, and aftringent medicines. For Afthma or Confumption has always followed, though poffibly at fome diftance, the injudicious treatment of eryfipelas, fmall-pox, meafles, miliary eruptions; fcorbutic fpots or puftules imperfectly expelled, or driven inward by the bark, healing up ulcers, repreffing fetid perfpiration of the feet, or obftructing any of the natural evacuations of the fyftem.

In the deftruction of *this Hydra with a thoufand heads,* the Vegetable Syrup is fimple and uniform in its operation, and acts by one power. It meets the fubtle, poifonous matter in the lymph or blood, however introduced; fits it to be difcharged, and affifts in difcharging it.

The patient in all thefe cafes is moft carefully to remember, when the medicine invites an efflorefcence or pimples to the furface, it frees the fluids from the impurities with which they are loaded; and that to repel them by cold, by the bark, &c. may be fatal.

CASE

C A S E   XXVII.

C O N S U M P T I O N.

In a Letter from CAPTAIN DORSET,

To Mr. S W A I N S O N.

Woolwich, January 25, 1787:

SIR,

I am sorry that indolence and forgetfulness have conspired to delay my writing a case in which your Syrup has been so efficacious. Your Pamphlet asked for such communications, and I determined to comply with the request, considering it less an act of justice to yourself, than of humanity to any who may labour under similar complaints. In the autumn of 1783, I was attacked by a disorder of the breast, which had all the symptoms of a pulmonary Consumption in its earliest stage; during the winter it grew worse; a considerable expectoration, (sometimes streaked with blood) hectic fever, and great wasting away ensued; I had the ablest medical assistance, and for upwards of twelve months adhered rigidly to a vegetable diet, asses milk twice a day, riding, change of air, blisters, caustics, &c. in short every thing that the most approved practice could direct; however, the disease still increased so fast, as to make it obvious I could not live the winter in England; accordingly a sea voyage, and the South of France, were recommended. The first took up six weeks, without being of the least service; the climate just kept me alive, but so much weaker, that in the summer I returned to England, merely to settle some business, proposing to go back to the Continent about September: but in the interim, the appearance of some

eruptions

eruptions on the fkin, together with a large abfcefs that had formed in my thigh, led me to fufpect fome fcorbutic tendency, and determined meto try your Medicine, which I had heard commended. The fuccefs exceeded all belief, nor can I defcribe its effects better than by faying, that in a fpace of time, almoft incredibly fhort, I was well, and grew fat; confequently all thoughts of going abroad were laid afide; and this is the fecond winter 1 am fpending without inconvenience, in a climate I before could not exift in. I do not choofe my name fhould " ftand rubrick on the walls," or figure in advertifements; but if it can have any weight in your private practice, I am fo convinced of the excellence of your medicine, that you are welcome to fhew my letter whenever you think it will recommend it.

I am, Sir,

Your moft obedient humble fervant,

M. DORSET.

C A S E   XXVIII.

C O N S U M P T I O N.

In a Letter from the Rev. DAVID WILLIAMS to the Author.

SIR,

Without the ufual introduction on gratitude, or public motives, to the influence of which I am not infenfible, —I will relate my cafe.

A fever left near my ancle a livid fpot; in the centre of which a puncture, almoft imperceptible, difcharged an

D 5   ichor

ichor which difcoloured my ftocking. I applied to a furgeon, who furnifhed me with a plaifter, and gave me medicines. The fpot taking a new appearance, and tending to an ulcer, I confulted a phyfician, who prefcribed pills, and ordered an iffue beneath my knee. Under that treatment my leg was healed; and on a journey in the following fummer, the iffue was dried up. I paffed the winter but indifferently; a cough, to which I had not been fubject, rendering the nights reftlefs. In the fpring my face and head were covered with pimples, fo that I could not wear my hair. I put myfelf under the beft medical direction in my knowledge; and ufed medicine internal and external, with various fuccefs, more than a year. I accidentally faw, in an old magazine, a preparation of brimftone ftrongly recommended in cafes like mine; and I determined on a trial of it. In a few weeks the pimples difappeared; but brimftone is a violent medicine; it brought on a dyfentery, from which I very narrowly efcaped, after a confinement of three months.

Under thefe proceffes, confumptive fymptoms gradually gained ground; and I could perceive my medical acquaintance fought only to check them by regulations of diet, or change of air.

My engagements being numerous, fome of them requiring the vigour and fortitude of the beft health, I was harraffed by the alternate neceffity of occafional abfence, and immoderate application. In this ftate, and without any caufe in my knowledge, I awoke in a morning nearly blind; my eyes and the regions around them inflamed; fiery fparks or corrufcations attending the admiffion of light, and rendering me totally incapable of bufinefs.

It would be tedious and diftreffing to the reader, to follow me through all the means I ufed to attempt the removal of that calamity, or to provide for fubmiffion to it. Some idea may be conceived of the fituation, by
knowing,

knowing, though my occupation and amufement were letters, I paffed two years without reading even a news-paper; without eating animal food, or drinking fermented liquor. The difficulty of my cafe was that of reconciling the ufe of mercury for my eyes, with thofe ftrong fymptoms of Confumption which often menaced my life.

Being reconciled to the vegetable diet, and relying on the high reputation of its advifer, I determined on a journey, and to remain in the country the whole fummer. I returned in improved health; but my fight continued weak and imperfect. A phyfician of great fkill in dif-eafes of the eyes was warmly recommended; and I put myfelf under his care. He bled me freely; and ordered pills compounded of antimony and mercury; affuring me, they would not bring on the afthmatic or pulmonary evils, which were deemed my conftitutional diforders.

Whenever I took mercury, it flattered my hopes, but always to aggravate difappointment. The complaints in and around my eyes were abated; but I was fo emaciated and fhattered, that I could hold hardly any thing in my hands; the pulmonary diforders returned with increafed violence; and the medicines and regulations were ex-changed for the bark, copaiba, pills, and a tonic diet. The exchange reftored my fpirits in fome degree, and I walked out; but in attempting to crofs a ftreet with a little rapidity, a fuppreffion of all my powers fuddenly took place, which were reftored the inftant I touched the pavement by a fall. I confidered it as the prelude of fpeedy diffolution; and having confulted every man of peculiar fkill in my general diforder, I refolved not to mention the accident to my friends, and to fubmit quietly to my fate. In a few days, as I was fitting at table, I fell on the carpet; and was once more anxioufly and humanely attended by medical friends. But the meafures propofed not awakening any hopes, I turned my my attention to the beft mode of being difentangled from all engagements; drew around me thofe on whom I could depend in helplefs extremities; and awaited the ftroke of my laft enemy.

In this general ftate, but with confiderable variations of health, you found me, when you repeatedly called, on account of difficulties created by lawyers in the purchafe of Velnos' Vegetable Syrup.

The difeafe had produced one effect, for which I cannot account. I had reconciled myfelf to death; and had arranged the moft trifling papers in my ftudy: but I had a defire of filence or concealment refpecting the fits, which had the appearance of fhame; and never mentioned or referred to them. I defired you to bring me a bottle of the medicine from your predeceffor, for complaints in the urinary paffage, to which I was really fubject, and which were branches of the common evil: but my intention was to make an experiment on the difeafe. It had no effect; and if you had not obferved that you meant to prepare the Syrup with more care, I fhould not have made a fecond trial.

It will be fufficient for thofe who perufe this account to be informed—that the firft courfe under your direction, confifted only of four bottles; and that I took them without expectation of relief. But perceiving the pulmonary evil abating, and that the fits did not return, the love of life roufed my hopes; and I frequently retarded the general purpofe, by taking larger quantities than my ftomach would bear. I applied intenfely two years to fuit the tendencies or operations of the Medicine to the various effects of the general difeafe on my conftitution; and I took of it forty-eight bottles. I have been three years free from all pulmonary complaints; all inconvenient eruptions; all inflammations of the eyes; and all fits. My general ftate of health is as good as I have ever enjoyed;

enjoyed; and I have fuftained greater fatigue of bufinefs this winter, than in any three years fince I have been in London.

I fhall willingly attend to the perfonal applications or meffages of your patients, in circumftances fimilar to mine; and add my teftimony to that of your numerous and powerful friends, on the great honour and integrity with which you adminifter a very valuable medicine.

I am, Sir, moft fincerely yours,

DAVID WILLIAMS.

No. 28, Great-Ruffel-ftreet,
Bloomfbury, Feb. 26, 1789.

————————————————————

C A S E   XXIX.

A S T H M A.

MARY EGLINTON, daughter of Thomas Eglinton, No. 2, Little Peter-ftreet, Weftminfter, aged about eleven, fix months ago was afflicted by a fhortnefs of breath, and a fwelling of the abdomen.—On applying to Dr. Smith, Blackfriar's-Bridge, he faid it was the Evil in her bowels; and prefcribed fome medicines, which were taken for about a fortnight, without effect. Her fhortnefs of breath was fo great, that fhe could not walk acrofs the room, and her belly was fwollen to an enormous fize. In this ftate fhe was recommended to take the Syrup, which entirely removed her complaints. Mr. Eglinton had two children who died of the fame difeafe.

PALSY.

MORBID humours, whether introduced by accidents, fevers, the small-pox, the measles, or by a more disreputable disease, if deposited on the skin, produce pimples, heats, and itchings; if carried to the lungs, inflammation, consumption, or asthma; if to the bowels, dangerous colics, cramps, hypochondriac complaints; if to the liver, jaundice; if to the head, convulsion, epilepsy, palsy, and madness.

Though the Syrup of Mr. De Velnos is peculiarly calculated to correct humours; is a mild and powerful diaphoretic; and expels morbific matter, by almost all modes of evacuation; Mr. Swainson would not have ventured to advise it in Palsies, if not warranted by such cases as the following.

C A S E   XXX.

P A L S Y.

Mrs. SWAINSON, No. 12, Great May's Buildings, St. Martin's Lane, in the winter of the year 1782, had a stroke of the Palsy; it nearly took away the use of her

her left side, particularly her arm, which seemed totally
dead. An apothecary of great skill and reputation ordered the arm to be rubbed with a blistering ointment,
which inflamed it to a great degree; and it remained
many days in that shocking state, notwithstanding the
inceffant application of fomentations and poultices.
Some scorbutic symptoms appearing about her, Mr.
Swainfon recollected, that some years before, she had
been relieved of eruptions and pimples by Velnos' Vegetable Syrup, for which she had taken great quantities of
drops and diet drinks in vain. Mr. Swainfon had hopes
that the Syrup might also abate and remove the inflammation in the arm; but not the slightest idea that the
Palfy would be affected. To the great aftonifhment and
joy of the whole family, a warmth was felt through the
difeafed fide, which for many weeks had been as cold as
ice. The tranfports of a condemned criminal, on re-
ceiving pardon, could not be greater than her's; for she
had been left hopelefs by the faculty. In about three
months she was perfectly cured, and reftored to a better
ftate of health than she had enjoyed for many years.

As this event determined Mr. Swainfon not to become
a quack doctor, (for he will never fink the uniform
reputation of his life to act in that fufpicious and injurious capacity) but to purchafe, at a very high price, the
original receipt of Mr. De Velnos, and to pledge himfelf
to fee it prepared and fold with the ftrictest integrity
and honour, it is his happinefs that he can in addition to
his own, and that of Mrs. Swainfon, refer to the teftimony of all his friends, and almoft all his acquaintance,
for the truth of this extraordinary event in his family.

.

CASE

C A S E   XXXI.

MR. JOHN FARQUARSON, No. 51, Queen-Anne-ftreet Eaft, near Portland Chapel, had a paralytic ftroke in the year 1780, which took away the ufe of his left fide. He had the advice and prefcription of two eminent phyficians; but the diforder baffled their fkill. The cafe of Mrs. Swainfon being known, he was advifed to apply in Frith-ftreet. The diforder had remained upon him three years, and his age was beyond fixty; he was, however, cured by the Vegetable Syrup of Mr. De Velnos, and is now in very good health.

C A S E   XXXII.

To MR. S W A I N S O N.

Navigation-Office, Birmingham, Sept. 23, 1783.

SIR,

I had lain under a violent rheumatic pain in my head for a confiderable time, when about eight years ago I was advifed to the cold bath; the firft immerfion relieved me from the pain, but brought on a paralytic complaint, which for a fhort time deprived me of my mental faculties, the ufe of my left fide, and the fight of that eye; my faculties, and the fight of the eye, were in a few months reftored by medical affiftance, electricity, and fea-bathing; the left fide continued very weak
and

and helpless; the feverity of the two laft winters, or some other caufe unknown, very much impaired my remaining health and ftrength, infomuch that I was apprehenfive of a deep confumption, and the near approach of my diffolution. My left leg fwelled very much, and became difcoloured and very painful near the ancle; my left eye extremely weak; my breath fhort and fœtid; and my fpirits low. In this fituation it was natural to wifh for relief, and I was prevailed upon to take a few bottles of your Syrup, although the fpring was fo very cold and fevere: before I had taken four bottles, the fwelling and pain of my leg were removed; the circulation of my left fide and the ufe of it much increafed, and my left eye very much ftrengthened; my breath became fweet, and my fpirits lively and chearful. During the whole time I continued taking your Syrup, viz. from February to the latter end of April, I perceived my health, ftrength and fpirits, gradually increafe; and can with great pleafure, fatisfaction, and gratitude, fay, that in thefe refpects I never was better. My left fide ftill continues rather weak, and the perfect ufe is not quite reftored, but hope a few bottles, which I intend to take next fpring, will, with the bleffing of GOD, quite reftore it; and I doubt not but it would now have been perfect, had the winter and fpring been lefs fevere.

I am, Sir,

With the trueft refpect,

Your grateful humble fervant,

JOHN RIDYARD.

CASE

C A S E   XXXIII.

Edward Tighe, Esq; Member in the Irish Parliament for the County of Wicklow, having, in many extraordinary Cases, seen the Effects of the Vegetable Syrup, directed Henry Fox, my late Agent in Dublin, to try its Efficacy in the following Case.

MR. PHILLIPS, a farrier at Donnybrook, near Dublin, at the age of forty, had received several strokes of the palsy; had received the best advice, and taken great quantities of medicine, about fourteen months before Mr. Fox put him on the Vegetable Syrup, in the spring of the year 1787, which not only took away the use of all his limbs, but affected his reason, so that he could not distinguish his wife, or his nearest friends; and on being asked questions of the most familiar nature, he seemed unable to conceive, as well as to utter, any answer to them.

His head, legs, and indeed his whole body, were greatly swollen; and he was as helpless as a child.

The first effects of the Medicine were on the swelling; then on the understanding, and gradually on the limbs.

He took it in moderate doses; about the fourth day, his water, of which he made a considerable quantity, was black as ink, and so fœtid, that it could not be endured in the room.

He threw up clotted matter, like bits of liver; and ulcers were produced in his legs, which discharged profusely, until the offending matter was expelled by the various evacuations; his appetite, health, spirits, and understanding returned; and he carries on his business of smith and farrier as usual.

CASE XXXIV.

Under the Society mentioned above: and taken down by a Physician, August 16, 1786.

———— RIGBY, discharged from the Queen's Regiment of Dragoons about Christmas, on account of a paralytic affection of the left side, which rendered him incapable of doing any thing; and was afterwards made a pensioner at the Hospital at Chelsea. His speech and eye-sight were much affected; and he could not use his left hand. Began taking the Syrup, August 10, and is now much mended.

The appearances in favour of his recovering, were so promising, that he set off in a species of joyful intoxication, to visit his friends in some distant part of the country, where he remains.

WHITE

WHITE SWELLING.

C A S E   XXXV.

Under the Direction of the Society, &c.

June 28, 1787.

JOHN ROCKET, about nine years of age, has several brothers and sisters, all of whom are healthy, is afflicted with a very large tumour of the right knee; apparently, from the hardness, an enlargement of the bone, and two ulcers on the outside of the knee. His mother, a widow, knows nothing of the cause; but says it was perceived accidentally some time ago.

Examined, July 18, better; the tumour reduced.

Examined, Nov. 7, general health better. Ulcer mended.

The boy was sent into a parish workhouse by one of the gentlemen of the Society; where medical abuses are among the distressing evils which call for the reforming hand of wisdom and humanity. From the operation of those abuses, Mr. Swainson was obliged to request the Society to discontinue the Medicine: but the mother has told him the child is well.

TAPE

CASE XXXVI.

Mr. Swainson submits the following Letter, which is from a lady of rank, to the consideration of his readers; but he will not affirm, until he has made further experiments, that the Syrup will expel Tape-Worms.

SIR,

MY dear girl has taken one bottle of your Medicine; ten days after she began to take it, a long worm, of above ten inches, came from her; she has mended very fast ever since, so much so, that every body is surprised to see her look so well. I shall be glad if you will send me another bottle, carefully packed.

I remain, Sir,

Your humble servant,

Aug. 8, 1787.

Respecting the general diseases of children, from worms in the intestines, he is assured from experience, that the Vegetable Syrup corrects the crude and viscid
matter

matter on which they are fupported; increafes the activity of the bile; ftimulates the inteftines; and promotes the expulfion of the worms with the corrupted
humours.

All the common and boafted remedies have mercury;
which brings on inteftinal inflammations, fatal to moft
children.

Weak teas of tanfy or wormwood may be ufed with
the Syrup.

INFLAMMATION OF
THE BLADDER.

MR. SWAINSON would apprife his readers, that
cafes of this kind frequently occur in Frith-ftreet;
occafioned by the fuppreffion of the menftrual or hemorrhoidal fluxes; by healing up old ulcers in perfons
called fcorbutic; by the repulfion of matter to be difcharged by perfpiration; or by the impurity of the blood
and juices. The Syrup relieves, without affuming new
pretenfions, by its general property of correcting acrimony,
acting as a gentle diaphoretic, and purifying the lymph
and blood. Scorbutic, hypochondriac, and old people,
fubject to this diforder, fhould often take the Vegetable
Syrup, to cleanfe the morbid mucus, or matter; and
after it fhould ufe milk in their diet.

CASE

CASE XXXVII.

Mr. BOCK, Taylor, Compton-ftreet, Soho, about 50 years of age, in the fummer of 1785, was attacked by a diforder in the Bladder, which for more than twelve months was attended with excruciating difficulties in voiding his urine, and great irritation and pain in the neighbouring parts. The urine was of a thick confiftence and clayifh colour, and the fmell extremely offenfive.

After taking the prefcriptions of feveral eminent phyficians without relief, and defpairing of finding benefit from medicine, he remained in the moft dreadful fituation from conftant pain, and want of reft, till the autumn of 1787, when he was advifed to try the Vegetable Syrup, which in a few weeks reftored him to health. He now enjoys good health and fpirits.

ERYSIPELAS

ERYSIPELAS and FISTULA.

The Cafe of EDWARD TIGHE, efq; Member of Parliament for the County of Wicklow.

THOUGH the talents and merits of Mr. Tighe have been principally encouraged in Ireland, his education at Eton and Cambridge, his ftudy of the law in the Inner-Temple, his intimate connections and general acquaintance among the Englifh nobility and gentry, his acknowledged tafte, vivacity and wit, render the credit of his teftimony as extenfive as the efteem of his character and the praife of his virtues.

Mr. Tighe's firft fymptom of difeafed lymph was a *Fiftula in Ano*; on which two operations were performed in 1776. Soon afterwards eryfipelatous fores appeared; which, on being healed in one part, broke out in another; and fuccefsively occupied the whole body. From 1776 to 1783, under the directions of the firft phyficians and furgeons of the age, he tried every fcorbutic medicine in the *Materia Medica*. The prefcriptions which gave him flight and temporary hopes, were bark and mercury in conjunction; elm bark, mineral waters, farfaparilla, hemlock, &c. The moft effectual of thofe prefcriptions was bark and mercury in conjunction. It was recommended by a fkilful furgeon now deceafed, and approved by Sir Cæfar Hawkins, Mr. Tighe's particular friend. But while fome of the ulcers were healed by it, a fever and

and fomething like delirium enfued, and prevented a completion of the cure. Mr. Tighe mentioned this circumftance, with his ufual fagacity and judgment in favour of the Vegetable Syrup, which not only healed the ulcers, fiftula, &c. but purified the whole fyftem, without having fuch effects in the flighteft degree: a demonftration that it contains no mercury. In the feven years of mifery he endured, the difeafe gradually fpread dreadful ulcers on his arms, thighs, fcrotum, anus, &c. notwithftanding the moft fkilful and humane attendance of Sir George Baker, Dr. Hallifax, Sir Cæfar Hawkins, Mr. Charles Hawkins, Mr. John Hunter, late Mr. Ditcher, Mr. Walker, &c.

On finding the Vegetable Syrup agree with him, he took twenty bottles in twenty weeks; and not only got rid of his ulcers, fiftula, &c. but, to ufe his own phrafe, " he felt a hogfhead of new life poured into him," and at forty-eight he has the appearance of the health and vigour of five and thirty.

From the year 1783, when the cure was performed, to this time (1790) his affiduities have been unceafing in favour of the Vegetable Syrup; and I take this opportunity of gratefully acknowledging them, and expreffing my peculiar pleafure in refcuing men of fuch talents and virtues as thofe of Mr. Tighe.

E

PILES

PILES and FISTULA.

C A S E   XXXIX.

BENJAMIN DUTTON, Conflitution, Bedford-ftreet, Covent-Garden, about the end of September 1786, perceived a fwelling in ano, which he imagined to be the Piles. The fwelling increafed in a fhort time to an uncommon fize; which in about feven or eight days broke, and difcharged a great deal of matter for that day and night, and then ftopped, gathered, and difcharged again; which it repeated three times. Being reduced by the diforder, a furgeon was fent for, who pronounced it a Fiftula, and that an operation niuft be performed; which was immediately complied with, and gave him great eafe. It was dreffed with dry lint. The furgeon faid it might be long in healing, and advifed him to go to the hofpital. It difcharged a green offenfive matter. He confulted Mr. Swainfon, who put him on a courfe of the Syrup; and about Chriftmas, a perfect cure was compleated, and the patient enjoys a good ftate of health.

INFLAMMATION of the EYES,

AND

LOSS of SIGHT.

C A S E   XL.

IN the beginning of the year 1787, William Lewis, fon of Samuel Lewis, had a violent inflammation in the Eyes, and a confiderable eruption about the mouth. On taking him to St. George's Hofpital, the attending furgeons apprehended the lofs of one eye, if not both. Being only four years old, he was made an out-patient, and a blifter ordered behind the right ear. This produced no good effect, and the eruption fpread over his head, difcharging the moft offenfive matter in confiderable quantities. In three months the difeafe had proceeded in its ravages, wholly deprived him of fight in one eye, and daily threatened the lofs of the other, and he could take very little food; when his danger and mifery drew the attention of Wm. Pulteney, efq; M. P. who recommended him to Mr. Swainfon in the month of March of the fame year. Mr. Swainfon ordered the Vegetable Syrup of De Velnos in fmall quantities, on account of the reduced condition of the patient. In a fortnight he was fufficiently improved in health to take two table fpoonfuls night and morning. The medicine as ufual augmented the difeafed difcharge, reflored the natural fecretions, and the child proceeded rapidly in his recovery : his fight is perfectly

reftored,

reftored, his health re-eftablifhed, and he has no eruption on any part of his body.

Mr. Swainfon has the permiffion of Mr. Pulteney to add his teftimony and fanction to the account of this interefting cafe,

---

Mr. Swainson would alfo refer the anxious reader back to the cafe of Mr. Williams, in page 57, where a temporary blindnefs arofe from a fcorbutic indifpofition of the humours, though no difcharge took place at the eyes. The medicine acts in them as in all other cafes to which it is adapted, by correcting the acrid ferum; procuring a derivation to other emunctories; and ftrengthening the relaxed glands. Setons, iffues, and blifters, are generally ineffectual; very often attended with danger.

Patients of this defcription are fubject to deafnefs about the equinoxes, or in cloudy and rainy weather, from ferous humours and lax nerves.

Of all external remedies, vapours of the decoctions of bay berries, wormwood, camomile, fage, elder, or rofemary, are the moft innocent; but all do harm, unlefs the internal caufe be removed.

Obferve, whether any eruptions on the head have been prematurely dried.

CANCERS,

CANCERS, SCHIRROSITIES, INDURATIONS, &c.

C A S E XLI.

S C H I R R U S.

To Mr. SWAINSON.

Sir,

IN January 1778, while at New-York, I had a cold and sore throat for three weeks, and my left testicle was prodigiously enlarged and softened. For what reason I cannot guess, the disorder passed to the right; which continued three months in a similar state, but by degrees grew hard. I came to England in May, but did not apply for affistance till August, when a surgeon at Kingston-upon-Thames advised the fuspenfion of it in a trufs. I adopted this method for a year and a half; but it became fo painful, and was attended with fuch a general lofs of health, that I had recourfe to one of the moft eminent and humane furgeons in London, who pronounced it a Schirrous cafe, and advifed the extraction of it, as the only means of recovering my health; hinting it as his opinion, that it had arifen from fome unfortunate female connection. This, I knew, could not be the cafe. I determined not to fubmit to the dangerous operation he propofed; and he ordered me gentle phyfick, and camphorated fpirits as

an

an embrocation. After using the latter two months without relief, I applied, by his advice, a poultice of linseed meal, &c. for three months more: but having no prospect of a cure, I began to sink under the most melancholy despair. On applying a poultice of bread and milk for some months, it broke, and there issued a thin watery matter, of a dusky colour, which every day grew thicker and more offensive. I was then persuaded to consult a physician, who is now abroad; he ordered extract of hemlock internally, and a fomentation of hemlock and camomile. These dreadful medicines, after a long trial, afforded me no relief, and tended only to confirm the despair under which I must have sunk, if I had not heard of the astonishing effect of Velnos' Vegetable Syrup in recovering Mrs. Swainson of a palsy, occasioned by a scorbutic habit; and your determination in consequence to purchase the recipe; and to add the sanction of your character for judgment and integrity to its other recommendations to the notice and relief of your fellow creatures.

The state of the scrotum, and the enormous size and condition of the right testicle, you might better describe than I can. The inflammation, which had reached the abdomen, was very alarming, and threatened a mortification. I was bled, and took two doses of physic by your direction. I then took the Syrup, which in a few days totally removed the inflammation. The second bottle produced a copious discharge of matter; the swelling decreased; and it is impossible to express what I felt at the prospect which I had lost for five years, that my health and spirits would return. Before I had taken the fifth bottle my wounds were healed, and the diseased part restored to a state of perfect soundness and health. I took two bottles more by way of security; and having been recovered from a state of misery, wretchedness, and despair, by means of your Syrup, I think it my duty to thank you for the

attention

attention you paid me, and to intreat you will communicate my case to the world, that others in similar circumstances may experience the surprising virtues of the Syrup, and enjoy the relief and happiness which it has brought to me.

I shall take the greatest pleasure in answering the inquiries of any persons who may refer to me.

I am, Sir, with gratitude and respect,

Your much obliged, and

Most obedient humble servant,

ELLIS PRICE,

No. 48, Maiden-lane, Covent-garden.

Feb. 5, 1784.

Attested by

Thomas Mainwaring, apothecary, Strand;
Wm. Naylor, apothecary, Bedford-street, Covent-garden.

CASE XLII.

The CASE of J. MAKENZIE.

DESCRIBED BY DR. SMITH.

A middle-aged man, who had been for several years married, and had long been free from venereal complaints, felt a rheumatic pain in his left side, after an exposure to cold in the month of May last year; which pain ascended and descended on the same side from the collar-bone to the hip, sometimes extended, and sometimes contracted within the compass of half-a-crown, when it was very violent, and attended with a continual remitting fever. In this
way

way the patient continued till the month of Auguſt; when he plainly obſerved, all on a ſudden, the pain go down from the ſide to his left teſticle, where it occaſioned an inflammation, which laſted a week, when it entirely left that and went to the right teſticle, which it inflamed in the ſame manner. This inflammation laſted longer, and was brought to ſuppuration by poultices; in which ſtate it continued three weeks, when the patient went into an hoſpital, where he felt ſharp rheumatic pain, with fever, which however, by the uſe of ſome pills, went off. Nothing but poultices were applied externally for a whole month, when the patient left the hoſpital, without being the leaſt better. The right teſticle was entirely bare, covered with a thick green ſlough, very offenſive, but no great diſcharge, loſs of ſtrength, and ſlight fever. Under theſe circumſtances, the Syrup was adminiſtered at firſt, in ſmall doſes, which were gradually increaſed; and a common detergent was applied externally; which treatment created a genuine pus, and perfectly cloſed the ſore.

C A S E   XLIII.

ACCOUNT OF MEDICAL PROCEEDINGS IN A CANCER.

In a Letter from a German Phyſician walking the Engliſh Hoſpitals, to a Friend in Germany.

SIR,

I came to England, as I propoſed, with a view to that information which the reputation of Engliſh literature promiſes. The hoſpitals in and near London firſt attracted my attention. Magnificence of ſtructure, and large revenues,

venues, highly elevated my expectations; and you may imagine my difappointment, on finding their medical management to be in general negligent, often unfkilful, and fometimes cruel.

That I may not appear to feek fhelter in general charges, I will ftate facts; which I think confiderably affect the medical character of England.

In the year 1787, in fome vifits I made at St. George's Hofpital, a patient, whofe name was Shailer, attracted my particular attention. He was the fervant of James Allarde, efq; in Charlotte-ftreet, Bloomfbury. The right tefticle was laid bare; the interior part of the fcrotum confumed, and a copious difcharge of pus took place. He had been fent to the hofpital to have the tefticle extirpated, which Mr. John Hunter had declared to be the only method of cure. The application of cauftics, previous to the operation, occafioned an inflammation, which extended to the left tefticle, and brought on a fimilar fuppuration. Having waited two months, the inflammation continued, and his life in imminent danger, the poor man was extremely defirous of applying to the proprietor of Velnos' Vegetable Syrup, by whom his life had been once faved. This excited my aftonifhment; and I was very attentive to a competition in a land of fcience between the efficacy of a noftrum, and medical fkill.

A philofopher of my own country had introduced me to Mr. Swainfon, the proprietor, as a man of tafte and letters; and had highly extolled the medicine from his own experience: but I had no idea that in fuch cafes it was fuppofed it could fucceed, after all the efforts of a celebrated hofpital.

But I was to learn the medical peculiarities of Englifh cuftoms. At the hofpital, all enquiries were checked by rudenefs. Mr. Swainfon very candidly invited me to attend the cafe, which I thought hopelefs: for both

tefticles

testicles were laid bare; the scrotum was nearly confumed; what remained was hard and contracted; and
the whole furface of the wound difcharged a fetid ichor.
The patient was emaciated, and had ftrong hectic fymptoms, the general prognoftics of death. I obferved Mr.
Swainfon was embarraffed by the frequent hemorrhages
from the corroded veffels; that he gave the Syrup in
fmall quantities; and thought the recovery doubtful.
But though a phyfician myfelf, I muft own there was a
minute attention and humanity in his conduct, which I
fhould have been glad to have feen in the medical practice.

I carefully accompanied him in his vifits to this patient. As the bleeding ceafed, the dofes were augmented;
and by the aid of common dreffings, and a nourifhing
diet, the patient recovered his ftrength; the wound
affumed a frefh and found appearance; and the indurated
fcrotum foftened.

In a month, to my great furprife, the patient left his
bed. What! faid I—Am I to ftudy in England under
the proprietor of a noftrum! who, though liberal and well
informed, feems to hold in contempt many of the medical
practices of his country?

An accident had nearly defeated him in the moment of
victory. The patient was convivial; and on getting out
of bed, he would rejoice with his friends. In pulling out
a cork from a bottle of wine, one of the veffels between
his tefticles burft; and occafioned an hemorrhage which
continued three hours. I accompanied Mr. Swainfon to
his affiftance. By the lofs of blood, the vital powers were
confiderably impaired; but by a judicious ufe of the medicine and diet, the patient was perfectly recovered. I had
not feen in any hofpital a medical event fo truly important!

I fhall make my reflections on a future occafion, and I
fubfcribe myfelf, Sir, Your moft humble fervant,
London, Aug. 24, 1788.   ——.

Jaundice,

Jaundice, or Icterus, is often complicated with ague, and with fchirrus of the liver.—If calculi are not formed in the hepatic ducts, the medicine will cure, by its general property of diffolving tenacious humours, opening obstructions, and promoting all fecretions.

Fainting, delirium, melancholy, mania or madnefs, are the frequent confequences of obftructions of the menfes; the fuppreffion of hemorrhoidal fluxes, from tranflations of humours to the head, perturbations of mind, the ufe of ftrong wines, mercury, and other medicines, which agitate the humours, and depreis the ftrength. Thefe diforders have gone off in hemorrhages, cutaneous eruptions, and ulcers, for which I nave been confulted; and they have enabled me to trace the origin of the general difeafe. Thefe circumftances do not feem to be fufficiently attended to, even by the phyficians, who moft fuccefsfully apply themfelves to fuch melancholy cafes.

The great analogy in the operations of all fpecies of *virus* in the lymphatic fyftem, would induce me to try the Vegetable Syrup in hydrophobia, or the madnefs occafioned by the bite of a dog. Mercury is at this time the moft fuccefsful medicine; and the Vegetable Syrup feems deftined to prove in all cafes, that mercury is ufelefs in the Materia Medica; and that all metals are unfuitable or injurious to the human ftomach.

The following cafe, though apparently extending the province of the Vegetable Syrup, is ftrictly within the defcription given of it, " an effectual medicine where morbid humours are repelled, retained, or introduced." When the external veffels are obftructed, the humours fettle either in glands to produce tumours or ulcers, in the lungs to produce afthma and confumption, or in the

bowels

bowels to produce diarrhæas, bloody-fluxes, or mortal inflammations. The Syrup, in a gentle, falutary, and *per-
manent* manner, opens the excretory veffels, and relieves
difeafes which, though apparently various, are only fymptoms of fcurvy.

C A S E   XLIV.

BLOODY FLUX and HABITUAL DIARRHÆA.

——— SUART, efq; an eminent merchant in Lancaf-
ter, in confequence of a fevere cold, had a bloody-flux and
diarrhæa, which was treated in the ufual manner by the
medical gentlemen of the place, but was little affected by
their remedies. Alarmed at the probable confequences,
he had the beft advice in London; but with no better
effect. He continued for *feven years,* under the general
neceffity of feeking every half hour the convenience of a
motion. This rendered his life miferable, wafted his conftitution, and left him no profpect but a lingering and
untimely death. Abraham Rawlinfon, efq; member of
parliament for that borough, fufpecting that an acrimonious humour might be a caufe of the diforder, and having
feen the effects of the Vegetable Syrup, had the goodnefs
to accompany Mr. Suart to Frith-ftreet, to take the
opinion of Mr. Swainfon.

Mr. Suart was put on a courfe of the Vegetable Syrup
in the fpring of 1789, and in a fhort time was perfectly
cured. For the diforder has not returned; and from an
emaciated defponding condition, Mr. Suart is become fat,
healthy, and happy.

CONCLUSION.

WHEN the reader has attentively perused the preceding cases, he will easily comprehend and admit these supplementary observations.

In Female Cases, hardly any occurred in Frith-street, without the *Fluor Albus*; commonly called the Whites. —They proceed from the same vessels that yield the menstrual blood, and appear always in habits called scorbutic. Sterility, abortion, and many uterine diseases, are the consequences of this complaint.

The Vegetable Syrup has always succeeded; though, for reasons easily imagined, the cases cannot be publicly stated. It evacuates the serous humours in the first passages; corrects the indisposition of the blood; and promotes the natural excretions, which are always defective and irregular in this disorder.

Astringents, external and internal, in this, as in menstrual and other discharges, occasion tumours in the region of the pubes, dangerous and sometimes fatal.

Children at the breast have been brought with apthæ, or small pustules in the fauces, owing to indispositions in the milk of the nurses, to whom I give the medicine.

Children have likewise epilepsies and convulsions; not from scorbutic irritations, originating in themselves; but from the passions of their nurses; from the use (by the nurses) of spirituous liquors, mercury, or any acrid medicines.

F.   Hiccoughs

Hiccoughs often arife from obftructed perfpiration; from gouty or eryfipelatous humours repelled; from mercury, antimony, lead, &c.

Dyfenteries prevail greatly among my patients, particularly at the end of hot, dry, and clofe fummeis; when the animal juices are liquified, and difpofed to putrefaction. The fudden check to the perfpiration of thcfe corruptible parts, from expofure to cold air at night, or other caufes, gives rife to this difeafe. It is prevented by taking in the evening one fpoonful of the medicine.—Old; fcorbutic, who are generally confumptive perfons, are peculiarly liable to this diforder. The medicine acts on the fubtle malignant humours in the general mafs; and expels them by the cutaneous potes. By thefe means the fever is abated; and the afflux to the inteftines prevented. Clyfters, of the folution of gums, commonly given, are to be avoided, for they make ulcers foul; and by fuppreffing the flux, pen up the noxious humours, and heighten the danger of the difeafe.

Scorbutic patients are very fubject to inflammations of the ftomach, from arfenical poifons, which are adminiftered to them with impunity; virulent purgatives or emetics; mercurial, antimonial, and other metallic medicines; and particularly from the repulfion of acrid humours from the furface.

Hectics, flow fevers, dry coughs, fwelling of the belly, lefs of appetite and ftrength, wandering heats, and fudden tranfitions in the bowels from a loofe to a coftive ftate; —thefe are the general fymptoms of latent fcurvy; and the patients denominated confumptive, come to me generally on a milk diet.—I always order whey, and not milk; or milk prepared with a little manna, or conferve of rofes.

In hectics—there is always a difpofition to inflammation, ulcer, or fchirrus of the vifcera.—Balfams, lozenges, and all unctuous relaxants, do great harm, where the diforder

order proceeds from acrimony. The fame perfons are extremely fubject to bloody eruptions, or difcharges; in infants, from the nofe; in youth, from the lungs; in manhood, from the hemorrhoidal veſſels; and in age, from the urinary paſſages. The great object in all thefe cafes, is not to reſtrain the flux, but to correct the acrimonious humour which occafions it; and which, if it terminates in the corruption of any of the vifcera, is mortal.

The attentive reader will obferve, in the whole of this pamphlet, that the operation of Velnos' Vegetable Syrup is confined to obſtructions and injuries of the lymphatic and glandular fyſtem; and that its efficacy is on the difeafes which arife from thofe general caufes.

Infenfible Perfpiration is the moſt confiderable evacuation of the human body; according to fome medical writers, equal to half of what we eat and drink—if incompetent, the body is overcharged with acrid humours, which produce maladies on the ſkin; and when fixed on the interior organs, occafion difeafes of the moſt ſerious nature. The medicine removes the difeafes by correcting acrimony, and reftoring the infenfible perfpiration.

Whether the morbid humours be the effect of external obſtruction, or introduced by abforption, as in the ſmallpox, or in difeafes of a more difreputable nature, their injuries are on the lymphatic or glandular fyſtem; by depofiting themfelves in the mefentery, the lungs, the liver, or the head, they produce colic, confumption, jaundice, dropfy, or palfy;—and the efficacy of the Syrup in thefe difeafes, is not the boaſt of empyrical vanity, but a fact clearly accountable on principles which cannot be fuccefsfully difputed.

N. B. This pamphlet has been printed in a diſtinct form, by the defire of feveral families of the firſt confequence, who wifhed to withhold even the punifhments of
fome

fome vices from the infpection of delicate and uncorrupted youth. The Syrup of Mr. De Velnos owes its firft celebrity to its unrivalled efficacy in a difeafe, which once alarmed Europe, as if menacing the extirpation of the human fpecies. Mercury is the feeble barrier oppofed by the faculty to that dreadful evil, in which Velnos' Vegetable Syrup alone is a certain or infallible fpecific. Mr. Swainfon hefitates where he has any doubts; but in that diforder, no man ever underwent a courfe of his Syrup without obtaining a perfect cure.

Cafes of that nature are given in another pamphlet.

FINIS.

www.ingramcontent.com/pod-product-compliance
Lightning Source LLC
Chambersburg PA
CBHW021953190326
41519CB00009B/1238